Laboratory Manual for Physical Geology

Second Revised Printing

Emanuela Baer
Shoreline Community College

Eric Baer
Carla Whittington
Highline Community College

Copyright © 1999 by Eric Baer, Emanuela Baer and Carla Whittington

Revised Printing

ISBN 10: 0-7575-0455-8
ISBN 13: 978-0-7575-0455-6

Kendall/Hunt Publishing Company has the exclusive rights to reproduce this work,
to prepare derivative works from this work, to publicly distribute this work,
to publicly perform this work and to publicly display this work.

All rights reserved. No part of this publication may be reproduced,
stored in a retrieval system, or transmitted, in any form or by any
means, electronic, mechanical, photocopying, recording, or otherwise,
without the prior written permission of the copyright owner.

Printed in the United States of America
10 9 8 7 6 5 4

Preface and Acknowledgements

An Instructional Improvement Grant and the Office of the Dean of Instruction for Transfer Programs at Highline Community College supported the development of this lab manual. The authors appreciate their assistance. We also thank all of our students for trying and commenting on these labs while they were still being developed and providing suggestions for this and previous editions. Other geology instructors, especially Lisa Gilbert and James Loetterle have made significant contributions. Thank you.

About the Authors

Emanuela (Emma) A. Baer has a Laureate degree in Geology and a Doctorate in Earth Science from the University of Rome, "La Sapienza" in Italy, where she specialized in volcanology. She has traveled to and studied many volcanoes in the world, including Greece, Italy, the U.S., and Iceland. Emma enjoys hiking, rock-climbing, kayaking, cooking, and landscape photography. She has taught at Shoreline Community College since 2001.

Eric M. D. Baer has a Bachelors degree from Carleton College in Minnesota and a Ph.D. from the University of California, at Santa Barbara. He has studied volcanic rocks in Ecuador and Southern Japan. Eric enjoys backpacking, traveling, and eating. He has taught at Highline Community College since 1997.

Carla M. Whittington earned a Bachelors of Science from Indiana-Purdue University at Fort Wayne in 1993 and a Masters of Science from Indiana University in 1996. Her specialty area is igneous petrology. She completed an Environmental Manager Certificate with the University of Washington in 1998. Carla enjoys traveling, backpacking, and the volcanoes of the Cascades. She has taught at Highline Community College since 1998.

Note: The authors of this book will not receive royalties for its sale or adoption at Highline or Shoreline Community College

Table of Contents

1. The History of the Earth and Geologic Time   **1**

2. Dating Rocks   **7**

2.2 Relative Age and Dating Principles..........................9

2.3 Absolute Age and Radiometric Dating....................23

3. Mineral Properties and Identification   **35**

3.1 The Physical Properties of Minerals.......................39

3.2 Common Rock-Forming Minerals...........................49

4. Igneous Rocks   **59**

4.1 Igneous Rock Textures.......................................67

4.2 Igneous Rock Identification.................................73

5. Topographic Maps   **81**

5.1 Introduction to Topographic Maps and Contours.......85

5.2 Reading Topographic Maps..................................95

6. Sands   **99**

7. Sedimentary Rocks   **111**

7.1 Textures of Clastic Sedimentary Rocks.................117

7.2 Identification of Sedimentary Rocks.....................123

8. Metamorphic Rocks   **129**

9. Earthquakes   **143**

10. Density and Common Earth Materials   **157**

Lab 1 and 2

The History of the Earth & Geologic Time

In the late 1700s, James Hutton recognized that the Earth is very old, but for many years there was no reliable method to determine the age of the Earth or the dates of various important events in the geologic past.

Today we have a **geologic time scale** (Figure 1.1) that organizes all of Earth's history into blocks of time during which important events occurred. Geologists and paleontologists developed the geologic time scale during the eighteenth, nineteenth, and twentieth centuries. But why do they believe that the planet is 4.56 billion years old? How did they pinpoint the timing of the Earth's great events?

Early geologists were able to construct a chronological sequence of Earth's historic events by using basic scientific logic and an understanding of fossils and spatial relationships, such as where rock layers form. By examining the rock record and answering the questions "Which came first?" or "Which is younger?" the **relative ages** of rock units could be established. Relative age lists the order in which events occurred, but does not tell us the timing of those events in years.

In the twentieth century, radioactivity and the application of radiometric dating techniques have allowed late twentieth century geologists to determine the **absolute age** of rock units. Absolute age is age in years. Thus, absolute ages have been added to the chronological ordering of the Earth's great events to establish the modern geologic time scale.

In lab 1 you will do an exercise designed to acquaint you with the magnitude of geologic time. In lab 2 you will explore relative dating techniques and absolute dating techniques. Your instructor may assign one or both of them.

To Bring to Lab
lab manual   small metric ruler   calculator   pencil

Figure 2.1 The Geologic Time Scale

Eon	Era	Period		Epoch	Age (millions of years ago)
Phanerozoic	Cenozoic (Cz)	Quaternary (Q)		Recent or Holocene	
					0.01
				Pleistocene	
					1.6
		Tertiary (T)	Neogene (N)	Pliocene	
				Miocene	
					23.7
			Paleogene (Pε)	Oligocene	
				Eocene	
				Paleocene	
					65
	Mesozoic (Mz)	Cretaceous (K)			
					146
		Jurassic (J)			
					208
		Triassic (TR)			
					250
	Paleozoic (Pz)	Permian (P)			
					290
		Carboniferous (C)	Pennsylvanian (℗)		
					325
			Mississippian (M)		
					360
		Devonian (D)			
					410
		Silurian (S)			
					440
		Ordovician (O)			
					505
		Cambrian (□)			
					543
Precambrian	Proterozoic				
					2500
	Archean				
					3800
	Hadean				
					4560

Name _____

Pre-Lab 1

Read Lab 1 and answer the following questions. Always remember to show math work, if appropriate and always, always, always include units with your final answer.

1.   How many millions are there in a billion?

2.   In lab you will make a timeline 4.56 meters long to represent the 4.56 billion years of Earth's history:

   a) How long would 1 billion years be on the timeline?

   b) How many years would 100 cm represent?

   c) How many years would 1 cm represent?

3.   Draw a line that is 1 cm long.

Lab 1 The History of the Earth

Introduction

The Earth has changed dramatically and repeatedly over a history that spans nearly 5 billion years. Such immense spans of time are difficult for most of us to comprehend. They fall outside our range of human experience. We normally deal with much shorter time intervals, like the time of our next class or the number of days until the next test, or even the number of years until graduation!

It is important for students of geology to expand their sense of time. Extremely slow geologic processes, considered only in terms of human experience, have little meaning. To appreciate the magnitude of geologic time and the history of our incredible planet, you will be creating a timeline of important geologic events scaled to a size more tangible and familiar.

Instructions

1. Make a scaled timeline.
   You will be making a timeline of Earth's history on a long strip of adding machine tape. The timeline should be done to scale. A scaled representation requires that 10 cm on your timeline represent the same amount of time anywhere along the timeline and each amount of time, say 5 million years, be represented by the same distance throughout the timeline. To do this you will:

   a) Measure out a strip of adding machine tape 4.56 meters long. A meter stick will be provided in lab.

   b) Select one end of the tape to represent the present. Beginning at that end, mark off each billion years (1 billion, 2 billion, etc.)

   c) Starting with the oldest event (Event #1), mark off all of the important events in Earth's history shown in Figure 2.2. In each case you should write the date and event directly on the timeline.

2. Turn your timeline into your instructor on the date due.

Figure 1.2 Important Events in Earth's History

Event #	Date in years before present	Event
1	4.56 billion	Earth forms
2	4.4 billion	Oldest mineral grain found
3	4.1 billion	Oldest piece of rock ever found
4	3.9 billion	Oldest evidence of a continent
5	3.8 billion	First evidence of life
6	3.5 billion	First fossils (algae and bacteria)
7	1.8 billion	Free oxygen in atmosphere
8	1.1 billion	First fossil of a complex organism (a worm)
9	540 million	First abundant life found in the rock record
10	460 million	First fish
11	440 million	First land plants
12	410 million	First land animals
13	250 million	Largest mass extinction occurs
14	247 million	First dinosaurs
15	240 million	First mammals
16	220 million	Breakup of super-continent Pangaea begins
17	145 million	First flowering plants
18	65 million	Dinosaurs and other animals go extinct
19	30 million	Mammals/flowering plants become abundant
20	5 million	Beginning of Cascade Volcanic Arc
21	1.8 million	First primate in genus Homo
22	100,000	First Homo *sapiens*
23	13,000	Humans first inhabit North America
24	10,000	End of last Ice Age
25	8,000	Founding of Jericho, the first known city
26	2,000	Roman domination of the world
27	500	European rediscovery of the Americas
28	~34	Humans first explore the moon

(Please note that some of these ages may differ slightly from those given in your text or that you found in another source. These dates change, but the general order and rough position stay constant.)

Name _____

Pre-Lab 2.1

You should read the chapter about geologic time in your textbook and Lab 2.1. Answer the following questions.

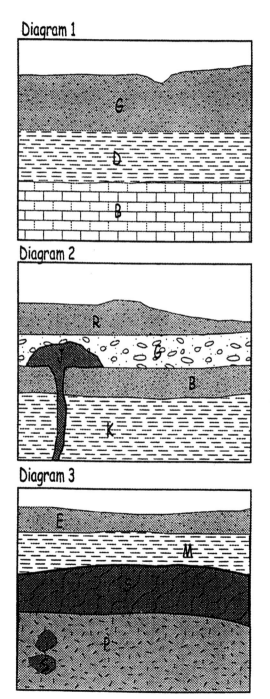

1. Based on the principle of superposition, which of the units in Diagram 1 is the youngest?

2. Based on the principle of cross-cutting relationships, which of the units in Diagram 2 is older: Unit G or Unit J?

3. Based on the principle of inclusions, which unit in Diagram 3 is older: Unit S or Unit P?

4. Name the type of unconformity between S and M in Diagram 3.

7

2.1 Relative Age and Dating Principles

Introduction

Relative dating allows geologists to interpret the geologic history of an area by placing geologic events in their proper sequence or order. Relative dating does not tell us how many years ago something occurred, only that something was preceded by one event and followed by another. For example, when looking at an outcrop you may notice two different sedimentary rock units. The lower unit might be a cross-bedded sandstone typical of wind deposits. The upper unit might be a conglomerate typical of a stream channel deposit. From this you might infer that the depositional environment in this location has changed from a desert environment to alluvial environment at some time in the past.

In this lab, you are shown four geologic cross-sections. These geologic cross-sections are side views of the rock beneath the surface in some hypothetical region. Each rock unit is assigned a specific pattern or symbol to denote its composition. A key to the rock symbols is provided in Figure 2.3. Using the information in the cross section, you will be asked to assign relative ages to the rock units, thereby interpreting the geologic history of each area.

Several logical principles can be used to help you establish the relative ages of rock units or geologic events:

> **Principle of Original Horizontality**: Sediments are deposited in layers that are horizontal or nearly horizontal. Therefore, when sedimentary rock layers are not horizontal, it can be assumed that they have been folded or tilted from their original position.

> **Principle of Superposition**: In any succession of sedimentary rock layers lying in their original horizontal position, the rocks at the bottom of the sequence are older than those lying above.

> **Principle of Cross-Cutting Relations**: Any geologic feature (intrusive igneous rock, fault, fracture, erosion surface, rock layer) is younger than any feature that it cuts.

> **Principle of Inclusions**: An inclusion in a rock is older than the rock containing it. Examples of inclusions are pebbles, cobbles, or boulders in a conglomerate, or **xenoliths** (pieces of other rocks) in an igneous intrusion.

Principle of Unconformities: An **unconformity** is an erosional surface that represents a gap in the geologic record. Unconformities are usually buried surfaces that mark a period of time in which deposition of sediments was interrupted due to uplift, deformation, and erosion. There are three types of unconformities. Each is identified based on the configuration of the rock units lying immediately below the unconformity (erosional) surface:

Nonconformities occur when igneous and/or metamorphic rocks lay beneath the erosional surface and sedimentary rocks lay above the erosion surface. Nonconformities often represent mountain building events in which uplift and deformation have occurred prior to continued sedimentation.

Angular Unconformities occur when tilted, folded, or deformed sedimentary rocks lay beneath the erosional surface and horizontally deposited sedimentary rocks lay above the erosional surface. Angular unconformities also represent mountain building events in which uplift and deformation have occurred prior to continued sedimentation.

Disconformities occur when horizontal sedimentary rocks lay both below and above the erosional surface. Although disconformities are erosional surfaces, there has been no deformation of the rock units during the non-deposition interval.

Figure 2.3 Key to Cross Section Symbols

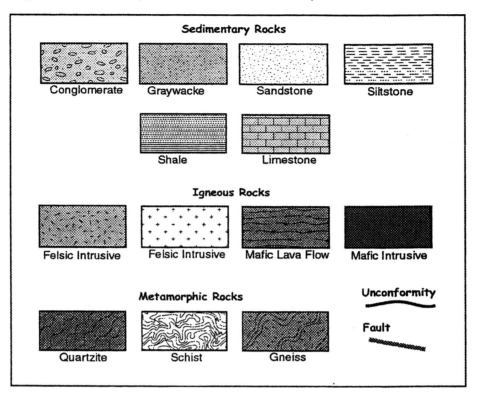

Figure 2.4 Example of a completed Cross Section

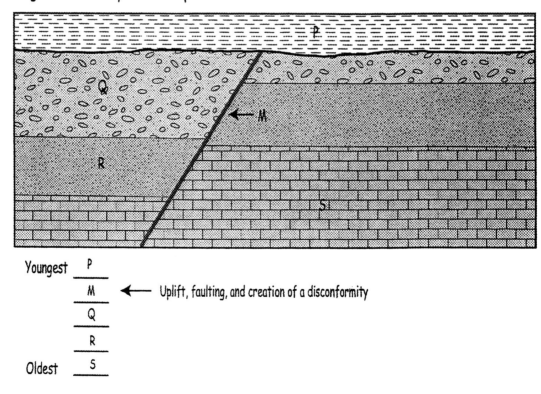

Instructions

1. Determine the relative ages of rock bodies and other features. Cross Sections 1-4 show hypothetical geologic sections through the Earth's crust. Apply the principles of original horizontality, superposition, cross-cutting relationships, and inclusions to determine the relative timing of the geologic events in each section.

   a) Indicate the relative age relations of the rock units and other lettered features by placing the letters on the blanks below each figure, from oldest (at the bottom) to youngest (at the top). Keep in mind the mode of origin of the various rock types and the principles for determining relative ages. Always start with the question: *What was there first?*

   b) Identify the periods of uplift, deformation, and/or erosion, by indicating along the sequence the timing of these events.

   c) Identify the type of unconformity created during the erosional event.

Name _____

CROSS SECTION 1

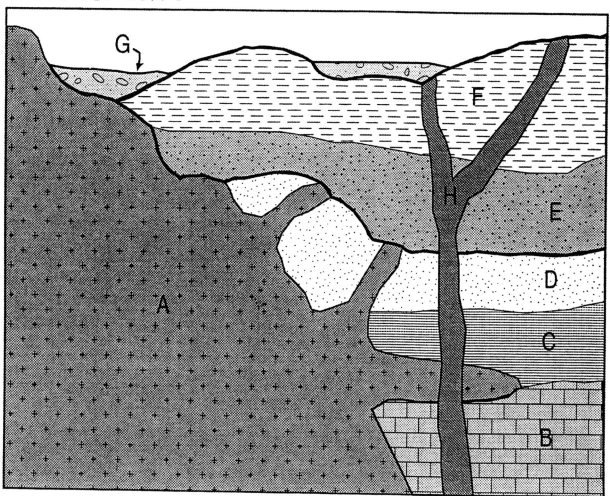

Youngest ____

____

____

____

____

____

____

Oldest ____

13

Name _____

CROSS SECTION 2

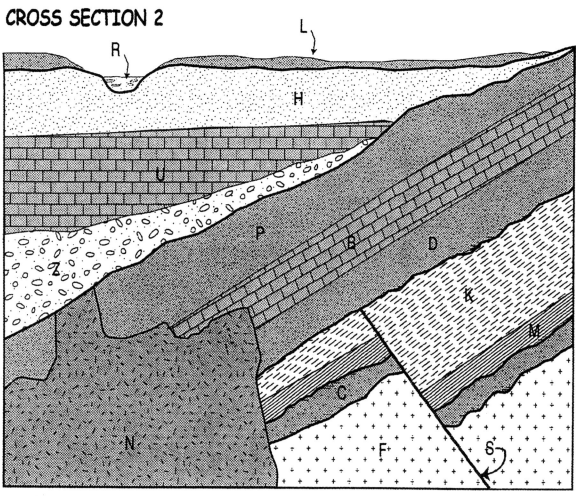

Youngest ____
____
____
____
____
____
____
____
____
____
____
____
Oldest ____

15

Name _____

CROSS SECTION 3

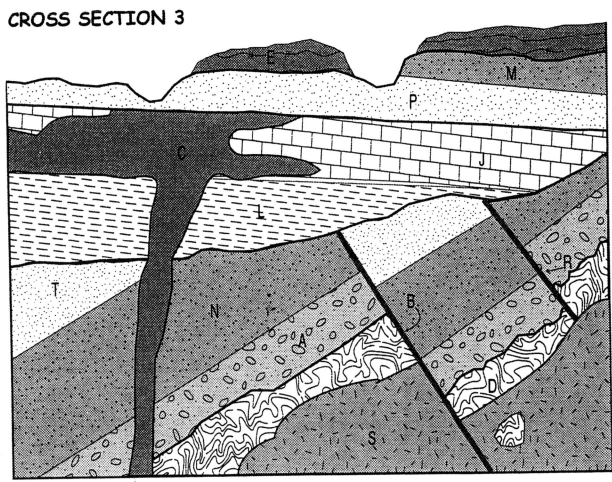

Youngest ___
___
___
___
___
___
___
___
___
___
___
___
Oldest ___

17

Name _____

CROSS SECTION 4

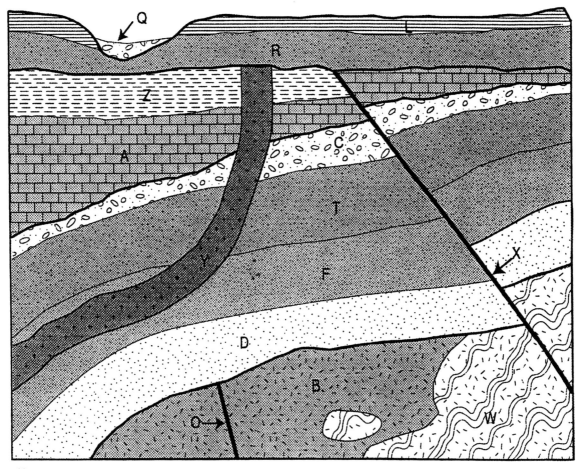

Youngest ____
____
____
____
____
____
____
____
____
____
____
____
Oldest ____

19

Name _____

Pre-Lab 2.2

Read the Introduction to Lab 2.2 (Absolute Age) and your book and answer the following questions.

1. What is an isotope?

2. What is the difference between a stable isotope and an unstable isotope?

3. What is the relationship between a parent and daughter isotope?

4. What is the half-life of a radioactive isotope?

2.2 Absolute Age & Radiometric Dating

Introduction

The discovery of **radioactivity** has provided a reliable means for calculating the **absolute age** in years of many of Earth's materials. Geologists calculate the absolute age of rocks using **radiometric dating** techniques. Radiometric dating relies on the spontaneous decay of the nuclei of unstable isotopes. These unstable isotopes, called **radioactive isotopes**, emit particles from their nuclei that we detect as radiation. The process of decay produces an atom that is stable and no longer radioactive. For example, ^{238}U (uranium 238) is a radioactive isotope. Its nucleus undergoes spontaneous decay and produces ^{206}Pb (lead 206), a stable isotope. The radioactive isotope (^{238}U) is referred to as the **parent isotope**. The stable isotope resulting from the decay of the parent is called the **daughter isotope**.

Radioactive decay is useful to geologists because it occurs at a predictable, measurable rate and is unaffected by geologic processes. The rate of decay is determined by the **half-life**, which is the amount of time needed for the original number of parent atoms to be reduced by one-half. For example, if we begin with 10 grams of radioactive material, after one half-life 5 grams would decay and become a daughter product. After the second half-life, one half of the remaining radioactive isotope, 2.5 grams or 1/4 of the original amount (1/2 of 1/2) would still exist. With each successive half-life, the remaining parent isotope would be reduced by half. The process of decay follows an exponential curve. With time the number of parent isotopes becomes very small, but since you can divide any number by two, even small ones, it never actually reaches zero.

In this lab, you will be creating a standard decay curve for a fictional element (Cascadium), calculating the half-life of this new and unusual element, and using the information from your graph to "date" rocks that contain the element.

23

Instructions ————————————————————————————————

1. Collecting the data.
   To create the decay curve, you will first need to collect data. Work in groups of 3-5. Start with a container of 50 atoms of Cascadium (represented by dice). We will assume that at time = 0, all atoms in the container are parent isotopes). Follow these steps:

   a) Shake the container and empty it onto a nearby tabletop or floor.
   b) Assume that any atom showing a one (1) is "Olympium", the daughter isotope of Cascadium. Remove the daughter isotopes, count them, and record the amount on the Decay Curve Data Sheet.
   c) Subtract the number of daughter removed from the number of parents remaining.
   d) Replace the Cascadium parent atoms in the container, shake and roll again.
   e) Repeat this process for 20 rolls or until there are no dice remaining in the container. Be sure to record the number of parent atoms remaining after each episode of shaking.

2. Create the decay curve.
   Plot your data on a piece of graph paper. Graph paper is provided at the end of your lab manual.

   a) For the horizontal (x) axis, you will plot the time in years. Assume time starts at 0 and each shake represents 2000 years.
   b) For the vertical article axis, you will plot the number of parent isotopes remaining. At time = 0, there will be fifty (50) parents.
   c) Be sure to label your axes and title your graph.
   d) After plotting the data, carefully draw a smooth, best-fit curve for the data.

3. Answer the questions that follow.

4. Answer the additional problems sections as assigned by your instructor.

5. Turn in your work (data table, decay curve, questions and additional problems) on the due date given by the instructor.

Name_____

Decay Curve Data Sheet

Shake #	Number of Years	# of Daughter Removed	# of Parents Remaining
0	0	0	50
1	2000		
2	4000		
3	6000		
4			
5			
6			
7			
8			
9			
10			
11			
12			
13			
14			
15			
16			
17			
18			
19			
20			

Name _____

Questions

Using the decay curve you created, provide the answers to the following questions. Show any math work you need to do and always, always, always include units on your final answer.

1.   Find the half-life of Cascadium from your graph.

2.   You have just found an unusual igneous rock and you want to know its age. You take the rock to an isotope lab for analysis. You are told that the rock contains only 16% of the original amount of Cascadium. The original amount was 50 atoms.

    a) Using your decay curve, find the age of this unusual rock.

    b) How many atoms of Olympium (the daughter isotope) are in the rock at the time of analysis?

3.   Analysis of a different rock shows that the ratio of parent to daughter isotope in the rock is 1/8 Cascadium to 7/8 Olympium. Calculate the age of this rock sample using the half-life determined in 1.

4.   If the rock in question 3 originally contained 14,000 atoms of Cascadium, how many are remaining after 4 half-lives have passed? How many Olympium atoms are there after 4 half-lives?

Name _____

Additional Problems Set 1

1. How much of the ^{238}U originally in the Earth is still present?

2. If you start with 1,000,000 radioactive parent atoms and their half-life is 1 week, how many parent atoms will remain at the end of 4 weeks?

3. What radiometric dating method (isotope pair) would be suitable for obtaining the age of each of the following?

    a) A rhyolite thought to be 1,000,000 years old?

    b) An Archean granite containing uraninite?

    c) Charcoal from an archaeological site thought to be about 10,000 years old?

4. A rock originally contained 8 grams of the radioactive isotope ^{40}K. At present only 1 gram of ^{40}K is in the rock. How many half-lives have gone by? How old is the rock?

5. What is the age of a grain of zircon that contains $^{235}U/^{207}Pb$ in the proportion of 1/16th parent to 15/16th daughter?

Name _____

Additional Problems Set 2

Using the following to complete the problems.

Dating System	Age of half-life
$^{238}U \rightarrow {}^{206}Pb$	4.5 billion years
$^{235}U \rightarrow {}^{207}Pb$	713 million years
$^{40}K \rightarrow {}^{40}Ar$	1.3 billion years
$^{14}C \rightarrow {}^{14}N$	5,730 years

1. A rock originally had 1000 atoms of ^{238}U and it now has 500. How old is it?

2. A rock originally had 2400 atoms of ^{40}K and no ^{40}Ar. It now has 200 atoms of ^{40}K. Assuming that the only chemical process is radioactive decay of potassium to argon, what is the age of the rock?

    a) 1.576 billion years
    b) 1.375 billion years
    c) 2.643 billion years
    d) 5.303 billion years
    e) 4.660 billion years

    How many atoms of ^{40}Ar are there in the rock now? _____

3. A rock originally has 1200 atoms of ^{235}U and no ^{207}Pb. It now has 220 atoms of ^{235}U. Assuming the only chemical process is the radioactive decay of uranium to lead, what is the age of the rock?

    a) 1.745 billion years
    b) 1.375 billion years
    c) 2.643 billion years
    d) 1.032 billion years
    e) 713 million years

    How many atoms of ^{207}Pb are there in the rock now? _____

31

Name _____

Additional Problems Set 3

The basic age equation is

$$N = N_o e^{-\lambda t}$$

Where N is the number of parents currently in the rock, N_o is the number of parents in the rock originally, e is 2.71...., t is the time that has passed and λ is the decay constant, which is related to the half life. In fact, the half-life (t1/2) is equal to $\ln 2/\lambda$. This equation is true for any exponential decay. However, since geologists usually want to solve for t, we rearrange the terms to

$$t = \frac{1}{\lambda} \ln\left(1 + \frac{D}{N}\right)$$

Where D is the number of daughter products today (N_o-N) and N is still the number of parent isotopes today.

Use the age equation to answer the following questions:

1. The half life of ^{235}U is 713 million years. What is λ?

2. Calculate the half life of an element with λ =.0000693

3. If you measure a rock with a ratio of daughter to parent (D/P) of 6, what is its age, assuming λ =.0693

4. You measure that a rock has 1,456,000 atoms of ^{40}K and 3,980,000 atoms of ^{40}Ar. How old is it? The half life is 1.3 billion years.

Lab 3

Mineral Properties and Identification

Introduction

We live in a world dependent on minerals. Without them, it would be impossible to live the way we do. Minerals are mined for aluminum to make beverage containers or copper to make electrical wiring. Minerals are used to make fertilizers to grow crops and perk up our lawns. We use minerals for manufacturing the products used to build our homes or run our computers. In fact, almost every manufactured product contains materials obtained from minerals. Fortunately for us, the Earth's crust is the source of a wide variety of essential and useful minerals.

How does a geologist identify a mineral in the field? Each mineral has a distinct chemical composition and crystal structure that distinguish it from all other minerals. For example, the mineral fluorite always consists of calcium and fluorine in a one-to-two ratio. But if you pick up a crystal of fluorite, you cannot see the ions. Measuring its chemical composition or crystal structure using laboratory procedures can identify the sample, but these analyses are time-consuming and very expensive. Instead, geologists routinely identify minerals by visual inspection of some common physical properties and by performing a few quick, simple tests.

The first part of this lab (3.1, The Physical Properties of Minerals) is designed to help you become familiar with the more distinct and common properties of minerals. You will learn how to make rather simple observations or perform simple tests to evaluate a particular property of minerals. You will then be given the chance to identify several minerals based on that property.

A few minerals are so distinctive that making a single observation or using a simple test can readily identify them. Unfortunately, this is not true of most minerals that you will encounter in your explorations. More often, you must make a combination of 4 or 5 observations and tests before an identification can be made. In the second part of this lab (3.2) you will be given several of the most common rock-forming minerals. You will use the tests you learned in part 3.1 to identify these minerals.

Name _____

Pre-Lab 3.1

Read the chapter on minerals in your text. Pay close attention to the characteristic properties of minerals.

1.   What is a mineral?

2.   Name five physical properties (described in section 3.1) that you can test to help you identify a mineral name.

_____   _____

_____   _____

_____

3.   Name five minerals.

_____   _____

_____   _____

_____

4.   Refer to the Alphabetic Mineral Index (Table 3.1) at the end of this chapter to answer the following questions.

   a)   What is the cleavage of gypsum?

   b)   What is the hardness of quartz?

   c)   What mineral has a hardness of 2-2.5, is black to brown, and has one perfect plane of cleavage?

37

3.1 The Physical Properties of Minerals

Instructions ———————————————————————

1.   Visit lab stations.

Your instructor has set up stations in the lab to illustrate different mineral properties. You may visit the stations in any order. At each station, you will:

    a)   Review the information on the physical property you will be evaluating or testing.

    b)   Record the sample number of the minerals at each station in the Mineral Identification Chart for Lab 3.1.

    c)   Perform the observation/test on each sample and record the results in the chart under the appropriate sample number.

    d)   Use the descriptions in the Alphabetic Mineral Index (Table 3.1) to make an identification of each sample.

    e)   Record the mineral name in the Mineral Identification Chart.

2.   Turn in your work on the due date given by your instructor.

Station 1:   Color and Streak

Color is among the more obvious qualities of a mineral, yet the color of a mineral may vary considerably depending on slight variations in chemical composition. Some chemical elements can create strong color effects, even when they are present only as trace impurities. For example, the mineral corundum is commonly white or grayish, but when small amounts of chromium are present, corundum is deep red and given the gem name *ruby*. Similarly, when small amounts of iron and titanium are present, corundum is deep blue, producing the gem *sapphire*.

Streak is the color of the fine powder of a mineral. Streak is observed by rubbing the mineral across a piece of unglazed porcelain known as a streak plate. Many minerals leave a streak of powder with a diagnostic color. Thus, streak is commonly more reliable than the color of the mineral itself.

The mineral names are provided at the station.

Station 2:   Luster

Luster is a property that describes the way light reflects from a fresh surface of the mineral. Minerals that have the appearance of metals, regardless of color, are said to have a *metallic luster*. Minerals with a *nonmetallic luster* are described by various adjectives, including glassy (vitreous), milky, and earthy (dull).

The mineral names are provided at the station.

Station 3:   Hardness

Hardness is the resistance of a mineral to scratching. The property of hardness is governed by crystal structure and by the strength of the bonds between atoms. The stronger the bonds, the harder the mineral.

Minerals come in a wide range of hardnesses. To compare the hardnesses of minerals, geologists use the Mohs Hardness Scale, shown in Figure 3.1. On the Mohs scale, each mineral is harder than those with lower numbers on the scale, so 10 (diamond) is the hardest and 1 (talc) is the softest. The Mohs Hardness Scale shows that a mineral scratched by quartz but not by potassium feldspar has a hardness between 6 and 7. Because the minerals of the Mohs scale are not always handy, it is useful to know the hardness values of common objects (see Fig. 3.1).

Figure 3.1 THE MINERALS OF THE MOHS HARDNESS SCALE

Minerals of Mohs Hardness Scale	Common Objects
1. Talc	
2. Gypsum	Fingernail
3. Calcite	Copper Penny
4. Fluorite	
5. Apatite	Knife blade
	Glass plate
6. Potassium Feldspar	Steel file
7. Quartz	
8. Topaz	
9. Corundum	
10. Diamond	

To perform a hardness test, first see if you can scratch the mineral with your fingernail. If you can, the mineral is soft. If not, try to see if the mineral can scratch a glass plate. If it can't, the mineral has intermediate hardness between 2 and 5. If it can scratch glass, the mineral is hard. The easier it is to scratch the glass, the greater the difference between the hardness of the mineral and the glass.

The mineral names are provided at the station.

Station 4: Cleavage and Fracture

Cleavage is the tendency of some minerals to break along flat surfaces. The surfaces are planes of weaknesses in the mineral's crystalline structure. Some minerals, such as mica, have one set of parallel cleavage planes. Such minerals will repeatedly break into smaller and smaller pieces along that one plane of cleavage. Others minerals have two, three, or even four cleavage planes. Some minerals have excellent cleavage. For instance, you can peel sheet after sheet from a mica crystal. Others have poor cleavage. Minerals with no cleavage are said to **fracture**.

Figure 3.2 shows common cleavage patterns that cause minerals to break into preferred shapes. Use this figure to help you evaluate the cleavage of mineral samples. There are three observations you must make when evaluating cleavage:

1. The number of different (non-parallel) cleavage planes. (Cleavage planes are relatively flat surfaces that give off flashes of reflected light when the hand sample is rotated.)
2. The angle at which the planes intersect.
3. The quality of cleavage: excellent, perfect, good, fair, poor

Fracture is the pattern in which a mineral breaks <u>other than</u> along planes of cleavage. Many minerals fracture because they have no planes of weak bonds in their atomic structure. Fracture can be in characteristic shapes. *Conchoidal* fracture creates smooth, curved surfaces and is common in the mineral quartz. Some minerals break into splintery or fibrous fragments. Others fracture into irregular shapes.

The mineral names are provided at the station.

Station 5: Density

An important physical property of a mineral is how light or heavy it feels. The property that causes this difference is **density**, the mass per unit volume. Minerals with a high density, such as gold, have closely packed atoms. Minerals with a low density, such as ice, have loosely packed atoms. The density of minerals is often reported as **specific gravity** (S.G.), the density of a substance relative to that of an equal volume of water. Most common minerals have densities in the range of 2.5-3.0 g/cm³. Metallic minerals have higher densities. For example, gold (Au) has a density of 19.3 g/cm³; galena (PbS) about 7.5 g/cm³, silver (Ag) about 10.5 g/cm³ and copper (Cu) is 8.9 g/cm³. Density can be judged by holding (*hefting*) different minerals of similar size and comparing their weights. Heavier minerals have higher than average densities.

The mineral names are provided at the station.

Figure 3.2 Common cleavage patterns of minerals. (From: *Laboratory Manual in Physical Geology*, 4/E by Busch, © 1997. Reprinted by permission of Prentice-Hall, Inc., Upper Saddle River, NJ.)

Number of Cleavage Directions	Shapes that Crystal Breaks Into	Sketch	Illustration of Cleavage Directions
0 No cleavage, only fracture	Irregular masses with no flat surfaces		None
1	"Books" that split apart along flat sheets		
2 at 90°	Elongated form with rectangular cross sections (prisms) and parts of such forms		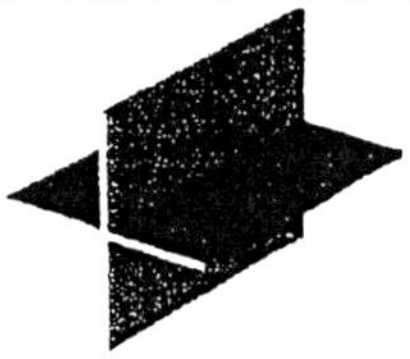
2 not at 90°	Elongated form with parallelogram cross sections (prisms) and parts of such forms	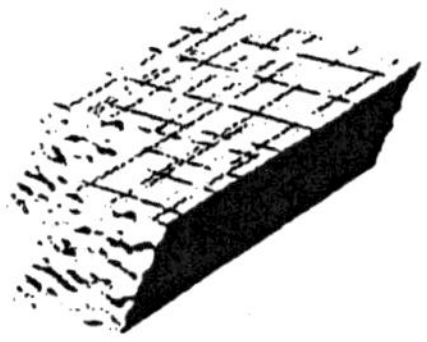	
3 at 90°	Shapes made of cubes and parts of cubes		
3 not at 90°	Shapes made of rhombohedrons and parts of rhombohedrons		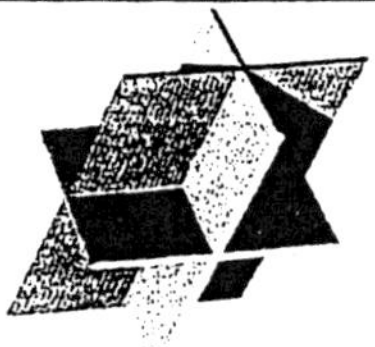
4	Shapes made of octahedrons and parts of octahedrons		
6	Shapes made of dodecahedrons and parts of dodecahedrons	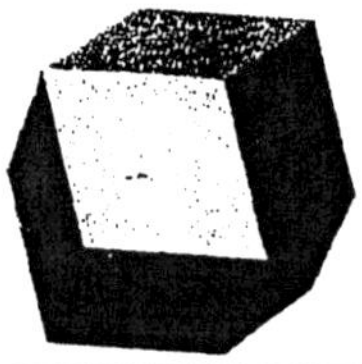	

Station 6:   Magnetism

Iron-bearing minerals often exhibit the property of **magnetism**. Some minerals are strongly magnetic, some weakly magnetic, and most are not magnetic at all. Magnetite is strongly attracted to a magnet; ilmenite and hematite exhibit a weak attraction.

Suspending a bar magnet on a string and slowly bringing the mineral in the vicinity of the suspended magnet is the quick test performed for magnetism.

The mineral names are provided at the station.

Station 7:   Reaction to Acid

Carbonate minerals (those containing the anion $(CO_3)^{2-}$) will **effervesce** (fizz) when a drop of dilute hydrochloric acid (HCl) is applied to a freshly exposed surface. The fizzing is the release of CO_2 gas, the same gas released when you pop the top of a soda bottle. Some minerals like calcite effervesce vigorously. Others, like dolomite, effervesce in dilute HCl only if the mineral is first powered.

You can easily perform an acid test by applying a small drop of dilute HCl to the mineral surface. If you get no reaction or a very slow reaction, you might want to scratch the mineral surface to form a powder, and then reapply the test. Remember to wipe the mineral dry after your test, so that the next person doesn't get acid all over their hands!

The mineral names are provided at the station.

Name _____

Please put sample numbers in numerical order.

Mineral Identification Chart for Lab 3.1

	Sample #	Color	Streak	Mineral Name
STATION 1				

	Sample #	Luster	Mineral Name
STATION 2			

	Sample #	Hardness Range	Mineral Name
STATION 3			

	Sample #	Cleavage or Fracture?	If Cleavage, # of Cleavage Planes	At 90° or not?	Mineral Name
STATION 4					

Mineral Identification Chart for Lab 3.1

	Sample #	Density (Heavy, moderate or light?)	Mineral Name
STATION 5			

	Sample #	Magnetism?	Weak or Strong?	Mineral Name
STATION 6				

	Sample #	Reaction to Acid?	Fast or Slow?	Mineral Name
STATION 7				

Name _____

Pre-Lab 3.2

You should use your book or Alphabetic Mineral Index (Table 3.1) at the end of this chapter for reference in answering the following questions.

1. Find the chemical formulas or compositions for all the minerals you looked at in Lab 3.1.

Mineral Name	Chemical Formula/Composition

2. Name eight minerals that are common in igneous rocks.

(Continued on next page)

3. Name 3 minerals you expect to be common at the Earth's surface and three

minerals that would be more rare there. Explain your answer.

Common   Rare

3.2 Common Rock-Forming Minerals

Instructions

1. Use the Mineral Identification Key (Table 3.2) to identify the minerals provided by your instructor.

   a) To use the Mineral Identification Key (Table 3.2), you must first determine the luster (*non-metallic* or *metallic*) of a mineral specimen. Minerals with non-metallic luster are further divided into *hard* or *soft* based on whether or not the mineral can scratch glass; then divided again on the basis of whether the mineral exhibits *cleavage* or *fracture*. Minerals that are metallic are divided into only two categories: those that exhibit cleavage and those that do not.

   b) In the Mineral Identification Chart for Lab 3.2 record the sample number of each mineral specimen. Evaluate or test the physical properties of the mineral and use the identification key to identify the minerals provided.

2. Turn in your work on the due date given by the instructor.

Name _____

Mineral Identification Chart for Lab 3.2

#	Luster	Hardness Range	Cleavage/ Fracture	Streak/ Color	Other Properties	MINERAL NAME

Mineral Identification Chart for Lab 3.2

#	Luster	Hardness Range	Cleavage/ Fracture	Streak/ Color	Other Properties	MINERAL NAME

Mineral Identification Chart for Lab 3.2

#	Luster	Hardness Range	Cleavage/ Fracture	Streak/ Color	Other Properties	MINERAL NAME

Table 3.1 Alphabetic Mineral Index

Mineral Name	Chemical Composition	Color	Streak	Luster	Hard-ness	Fracture or Cleavage	Other Properties	Geology, Uses
Amphibole	Hydrous Ca, Mg, Fe, Al-Silicate	Dark-green to black	Colorless to gray - green	Glassy on fresh surface	5 – 6	Good, 2 directions @ 56° and 124°		Important, common igneous rock forming mineral
Biotite	Fe, Mg, K, Al-Silicate (hydrous)	Dark brown to black	White to gray	Glassy	2 – 2.5	Perfect, 1 direction	Thin, elastic sheets	Common in igneous and metamorphic rocks
Calcite	$CaCO_3$	Usually colorless or white	White	Glassy	3	Perfect, 3 directions @ 75° (rhombus)	SG: 2.7 Fizzes in dilute acid	Major component of limestone, cement, contributes to hard water
Clay Minerals	Hydrous aluminum silicates	Various colors		Earthy, powdery	2 – 2.5	Irregular fracture		Used in ceramics, paper-making, kitty litter, food additive
Corundum	Al_2O_3	Gray, brown, red, blue	None	Opaque to glassy	9	Irregular fracture		Used as an abrasive and as gemstones (Ruby/Sapphire)
Dolomite	$CaMg(CO_3)_2$	White, pink, gray, brown	White	Glassy to milky	3.4 – 4	Good, 3 directions @ 75° (Rhombus)	SG: 2.85 Fizzes in dilute acid if powdered	Major component of dolomitic limestones and marbles
Fluorite	CaF_2	Clear, white, green, purple, yellow	White	Glassy	4	Good, 4 directions; not at 90°	Forms cubic crystals	Used as a flux in steel & aluminum manufacturing and in making hydrofluoric acid
Galena	PbS	Metallic gray	Gray	Metallic	2.5	Good, 3 directions @ 90°	SG: 7.5	Principal ore of lead
Geothite	$\propto FeO \cdot OH$	Yellow brown to dark brown	Yellow-brown	Earthy to glassy	5 – 5.5	Perfect, 1 direction though not exhibited in most specimens	SG: 4.37	Formed by oxidation of iron bearing minerals
Graphite	C	Gray or black	Gray or black	Metallic	1 - 2	Microscopic: 1 direction	Foliated masses, feels greasy	Used in pencils and as lubricants

53

Table 3.1 Alphabetic Mineral Index

Mineral Name	Chemical Composition	Color	Streak	Luster	Hard-ness	Fracture or Cleavage	z	Geology, Uses
Gypsum	$CaSO_4 \cdot 2H_2O$	Colorless or white	White	Glassy to milky	2	Good, in 1 direction, poor in other 2	Can be fibrous	Main ingredient of wallboard
Halite	$NaCl$	Colorless	White	Glassy	2.5	Perfect, 3 directions @ 90°	Tastes salty	Used as table salt, road salt and in chemical industry
Hematite	Fe_2O_3	Reddish brown to black	Reddish brown	Metallic to earthy	5.5 – 6.5	Irregular fracture	SG: 5:3	Coloring agent responsible for most red rocks. Major ore of iron.
Ilmenite	$FeTiO_3$	Black or dark-brown	Black to red brown	Metallic to earthy	5 – 6	Irregular fracture	Weakly magnetic	Major ore of titanium
Magnetite	Fe_3O_4	Dark gray to black	Black	Metallic	6	Uneven fracture	Strongly magnetic	An important source of iron
Muscovite	K, Al-Silicate (hydrous)	Colorless to light yellow/gold	White	Glassy to milky	2 – 2.5	Perfect, 1 direction	Transparent in thin sheets	Common in rocks, used as an insulating material
Olivine	$(Fe,Mg)_2SiO_4$	Olive to gray-green	Colorless to pale green	Glassy	6.5 – 7	Conchoidal fracture (often breaks around grains)	Usually granular, often massive	Important igneous rock mineral; gemstone variety: peridot
Plagioclase Feldspars	$NaAlSi_3O_8$ and $CaAl_2Si_2O_8$	White to dark gray, green, blue iridescent hues	White	Milky or trans-lucent	6	Good, 2 directions at or near 90°	Striations (grooves) often visible on cleavages	Important igneous rock mineral
Potassium Feldspar	$KAlSi_3O_8$ (Orthoclase or K-spar)	Pink, white, reddish	White	Milky or trans-lucent	6	Good, 2 directions at or near 90°		Important igneous rock mineral
Pyroxene	Ca, Mg, Fe, Al-Silicate	Dark green to black	Colorless to greenish gray	Glassy on fresh surface	5 – 6	Poor, 2 directions @ 90°	SG: 3.2	Important igneous rock mineral

Table 3.1 Alphabetic Mineral Index

Mineral Name	Chemical Composition	Color	Streak	Luster	Hard-ness	Fracture or Cleavage	Other Properties	Geology, Uses
Pyrite (Fool's Gold)	FeS_2	Brassy yellow	Greenish black to black	Metallic	6 – 6.5	Irregular or Conchoidal fracture	SG: 5:02 Forms cubic crystals	Source of much sulfide pollution and acid mine drainage
Quartz	SiO_2	Colorless or any color	Colorless	Glassy	7	Conchoidal fracture	SG: 2.65 hexagonal crystals	Common in rocks; varieties: Amethyst, Citrine, Rose Quartz, etc.
Sphalerite	ZnS	Yellow brown to black	Pale yellow, brown	Glassy (opaque)	3.5 – 4	Good, 6 directions	SG: 3.9 – 4.1 Smells like sulfur	Ore of zinc
Sulfur	S	Yellow	Pale yellow	Glassy	1.5 – 2.5	Conchoidal fracture	SG: 2.1	Principal source of sulfur

Table 3.2 Mineral Identification Key

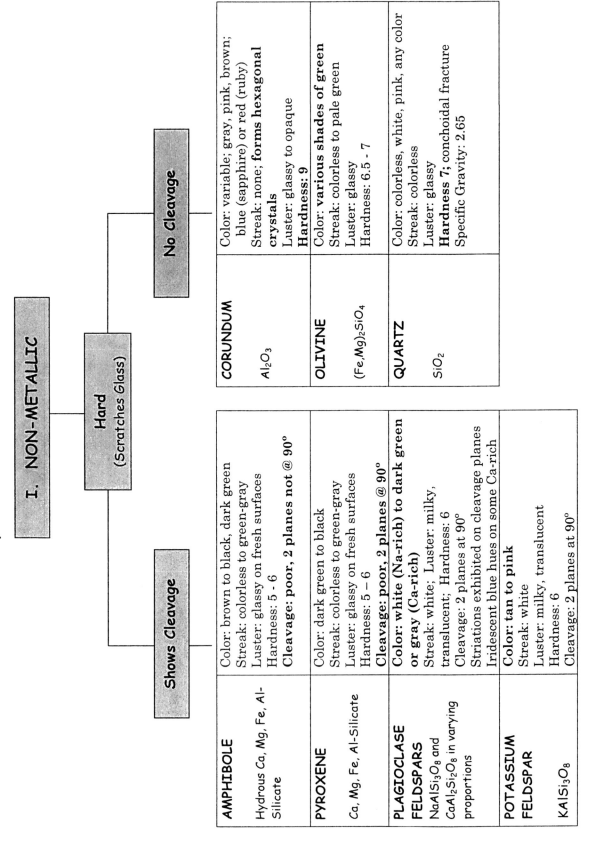

Table 3.2 Mineral Identification Key

II. NON-METALLIC

Soft (Does Not Scratch Glass)

No Cleavage

GOETHITE	Color: yellow brown to dark brown Luster: earthy to glassy **Streak: yellow-brown** Hardness: 5 - 5.5 Spec. Gravity 4.37
αFeO·OH	
HEMATITE	Color: varies from reddish brown to black Luster: earthy, dull, to submetallic **Streak: reddish brown** Hardness 5.5 – 6.5 Specific Gravity: 5.3
Fe_2O_3	
SULFUR	Color: yellow Luster: glassy; **Streak: pale yellow** Hardness: 1.5 – 2.5 Specific Gravity: 2.1
S	

Shows Cleavage

BIOTITE	Color: dark brown to black Streak: white to light black Luster: glassy to submetallic Hardness: 2 – 2.5; Cleavage: perfect, 1 plane; **Thin, elastic sheets**
Fe, Mg, K, Al-Silicate (hydrous)	
CALCITE	Color: colorless, white, yellow Streak: colorless or white Luster: glassy; Hardness: 3 Cleavage: 3 planes @ 75° **Effervesces in dilute HCl**
$CaCO_3$	
DOLOMITE	Color: white, pink, gray, brown Streak: white; Luster: glassy to milky Hardness: 3.4 – 4; Cleavage: 3 planes @ 75° **Powder effervesces in dilute HCl**
$CaMg(CO_3)_2$	
FLUORITE	Color: any color; usually clear, yellow, purple; Streak: white; Luster: glassy Hardness: 4; Forms cubic crystals **Cleavage: good, 4 planes not at 90°**
CaF_2	
HALITE	Color: colorless Streak: white, Luster: glassy Hardness: 2.5; **Salty taste** Cleavage: perfect, 3 planes @ 90°
NaCl	
GYPSUM	Color: colorless to white Streak: white; Luster: glassy to greasy; **Hardness: 2;** Tabular crystals, sometimes fibrous; Cleavage: perfect in 1 direction; poor in other two
$CaSO_4 \cdot 2H_2O$	
MUSCOVITE	Color: colorless to light yellow/gold Luster: glassy, milky; Streak: white Hardness: 2 – 2.5 Cleavage: perfect in 1 direction; **Thin, elastic sheets**
K, Al-Silicate (hydrous)	

Table 3.2 Mineral Identification Key

III. METALLIC LUSTER

Shows Cleavage

GALENA	Color: gray
PbS	**Streak: gray**; Hardness: 2.5
	Specific Gravity 7.5 (very dense)
	Cleavage: good, 3 planes @ 90°
SPHALERITE	Color: yellow-brown to black
ZnS	**Streak: pale yellow to yellow-brown**
	Hardness: 3.5 – 4
	Specific Gravity: 3.9 – 4.1
	Cleavage: good, 6 planes

No Cleavage

GRAPHITE	Color: gray to black
C (Carbon)	**Streak: gray or black**
	Hardness: 1 – 2; Scaly foliated masses
	Cleavage: microscopic does exist
HEMATITE	Color: reddish brown to black
Fe_2O_3	Metallic variety can be micaceous
	Streak: reddish brown
	Hardness 5.5 – 6.5
	Spec. Gravity: 5.3
ILMENITE	Color: dark brown to black
$FeTiO_3$	**Streak: black to reddish brown**
	Hardness: 6
	Weakly magnetic
MAGNETITE	Color: dark gray to black
Fe_3O_4	**Streak: black**
	Hardness: 6
	Strongly magnetic
PYRITE	Color: brassy yellow
(Fool's Gold)	**Streak: green-black to black**
FeS_2	Hardness: 6 – 6.5
	Specific Gravity: 5.02

58

Lab 4

Igneous Rocks

Introduction

In this lab you will learn to identify some of the more common igneous rocks. First you will learn how to distinguish between intrusive and extrusive rocks and then you will identify the major intrusive and extrusive rocks.

Igneous rocks form from cooling of magma or lava. There are two main classes of igneous rocks. If magma pools and cools slowly underground, it solidifies into an igneous **intrusive** (or **plutonic**) rock. If magma reaches the Earth's surface, and it is erupted by a volcano, it cools quickly and it solidifies into an igneous **extrusive** (or **volcanic**) rock. Extrusive rocks may form during different types of volcanic eruptions. Effusive eruptions of degassed magma (lava) form extrusive **effusive** rocks. Explosive eruptions, in which magma and gases are mixed together and violently ejected at the volcanic vent, will form **pyroclastic** rocks.

Textures of Igneous Rocks

Because of their different cooling histories, intrusive and extrusive rocks develop a different texture, which is the most distinctive feature for their identification.

Phaneritic texture

If magma cools slowly underground the elements present in magma have plenty of time to bond together to form the different minerals of an intrusive rock. As a result, an intrusive rock displays several well-developed and relatively large crystals, visible to the naked eye (larger than 1 mm). We call this texture **phaneritic.** Each mineral forms within its own temperature interval and the late-forming minerals fill the remaining spaces creating a mosaic of crystals.

Aphanitic texture

When magma is extruded on the surface of the Earth by a volcano, the sudden change in temperature causes very fast cooling which does not allow enough time for the elements to bond together and form large crystals. As a result, an extrusive rock rarely displays crystals visible to

59

the naked eye. The minerals may have formed but they are often too small to be seen without large magnification. We call this type of texture **aphanitic**. These rocks are usually very uniform in their appearance. Their color ranges from black to light gray or white, according to their chemical composition and the microscopic minerals present in the rock.

Porphyritic texture
There are some extrusive rocks in which a few large minerals are present (larger than 1 mm), embedded in a uniform **groundmass**, the much smaller (microscopic) mineral grains that usually make up the bulk of the rock. How could have these crystals grown so much if the cooling was so fast? The answer is that they did not form when magma was erupted at the surface of the Earth, but that they had already crystallized before the eruption, when magma was still pooling (and cooling) underground in the magma chamber. These early-formed minerals are called **phenocrysts** and the resulting texture of these types of extrusive rocks is called **porphyritic**.

Vesicular texture
Pyroclastic rocks are quite different from any other type of extrusive rock because they have **vesicles**, small holes left by the escaping gases during fast cooling. As a result these rocks tend to be much less dense than extrusive or intrusive rocks. Their color varies according to the chemical composition.

Glassy texture
Both effusive and explosive rocks may occasionally form from magma that cooled so quickly that that no minerals (not even microscopic ones) had time to form. These rocks are made of pure volcanic glass and their texture is called **glassy**. **Obsidian** is a notable example

Criteria for Identifying Igneous Rocks
Intrusive vs. extrusive rocks
Intrusive and extrusive rocks deriving from the same magma have identical chemical composition and even identical mineral content but their different cooling history will result into different textures. In order to distinguish between an intrusive and an extrusive rock you will need to look at the texture of the rock.

Intrusive rocks always have a phaneritic texture so it is easy to tell them apart from extrusive rocks. All that you see in intrusive rocks are minerals. Extrusive rocks may have an aphanitic, porphyritic, vesicular or glassy texture.

Intrusive rocks

To tell different types of intrusive rocks apart you need to look at the mineral composition and estimate the relative percent of each mineral in the rock. You can do this by using Figure 4.1 and the chart in Figure 4.2. The latter classifies igneous rocks on the basis of their texture, silica content, color and percent of minerals by volume. For each intrusive rock there is an equivalent extrusive rock with different texture but identical chemical composition.

Extrusive rocks

To distinguish among different types of extrusive rocks the charts in Figures 4.1 and 4.2 are not always useful. This is because extrusive rocks are often aphanitic and therefore the mineral content cannot be identified by looking at a hand specimen. In these cases, the color is a much more useful characteristic because it is dependent on the amount of silica and therefore on the amount of the **felsic** and **mafic** microscopic minerals in the rocks. Mafic rocks have a low silica content but they are rich in iron and magnesium. This gives them dark colors such as gray or black. Felsic rocks have high silica content but they are low in iron and magnesium. This gives them light colors, such as white, light gray and pale pink. Rocks with medium amounts of silica, iron and magnesium have intermediate colors such as various shades of gray. The presence and percentage of phenocrysts will be useful in porphyritic rocks in order to cross-check the color criterion.

A special type of effusive rock with a glassy texture is **obsidian**. Despite its high silica content obsidian is usually black. Because of the high silica content obsidian is very hard and it displays a very characteristic conchoidal fracture.

Effusive vs. explosive rocks

Effusive and explosive rocks may have identical chemical compositions. To tell them apart you need to look at the presence of vesicles and the density of the rock. Remember that explosive rocks are always vesicular and much less dense than effusive ones of the same composition.

Explosive rocks

In this lab we will look at two types of pyroclastic fragments, **pumice** and **scoria**, commonly found in explosive rocks.

Pumice is a fragment of hardened magma, erupted during very explosive eruptions with medium to high gas content in magma and medium to high silica content. Pumice is usually light-colored, white or gray and has many very small vesicles, which make it quite light. Pumice is less dense than water so it floats.

Scoria is a fragment of hardened magma erupted during mildly explosive eruptions with magmas low in silica and relatively low in gasses. Scoria is usually black or red in color (from oxidized iron) and moderately vesicular with fewer, but often larger vesicles than a pumice. Although scoria is less dense than an equivalent effusive rock, it is denser than water (and pumice) and sinks.

More than one texture could be applicable to a single explosive fragment. For example, pumice is always vesicular but is also generally glassy.

Table 4 summarizes the information necessary to identify and describe the various types of extrusive rocks.

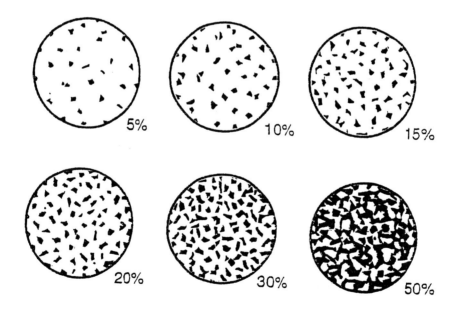

Fig. 4.1 Comparative diagrams to estimate percent of minerals in igneous intrusive rocks. (from "Learning in the Natural Laboratory", M. Russel Robertson, N. West and K. Stewart, pg 61, Kendall Hunt, 1997)

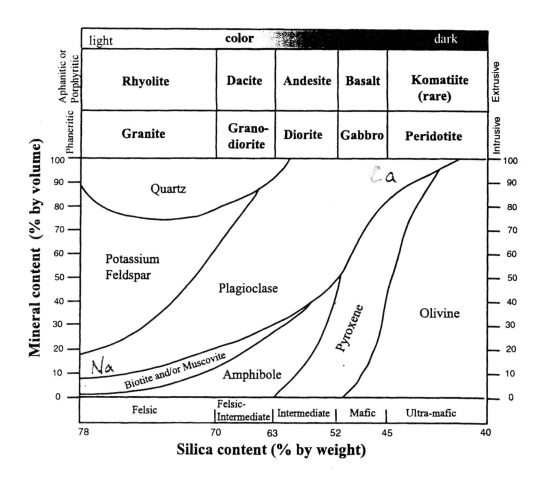

Fig. 4.2 Igneous Rock identification diagram.

Table 4: Extrusive Rocks Identification Key

Effusive rocks (from effusive eruptions)

Texture	Name of rock	Silica content	Color
Aphanitic, porphyritic	BASALT	Mafic	Black or reddish-brown
	ANDESITE	Intermediate	Dark gray
	DACITE	Felsic-intermediate	Gray
	RHYOLITE	Felsic	Pale gray, white or pink
Glassy	OBSIDIAN	Felsic	Red or Black

Pyroclastic rocks (from explosive eruptions)

Texture	Name of Rock
Vesicular and glassy	SCORIA
	PUMICE

Name: _____

Pre-Lab 4.1

Read the section on igneous rocks in your textbook and the introduction on igneous rocks in this lab manual. Answer the following questions

1. List 8 minerals that are common in igneous rocks.

2. How can you tell an igneous intrusive rock from an extrusive one?

3. If a rock has 15% quartz, 55% plagioclase, 5% potassium feldspar, and 5% biotite and 20% amphibole, what is its name?

4. What mineral is present in almost all types of igneous rocks (except sometimes ultra-mafic)?

4.1 Igneous Rock Properties

Instructions ———————————————

1. Visit lab stations.
   Your instructor has set up stations in the lab to illustrate different igneous rock textures and properties. You may visit the stations in any order. At each station you will:

   a) Review the information on the property you will be evaluating or testing.

   b) Record the sample number of the rocks at each station in Table. 4.1 (Igneous Rock Property Table).

   c) Perform the observations on each sample and record the results in Table 4.1 under the appropriate sample number.

2. Turn in your work (Table 4.1) on the due date given by your instructor.

Station 1: Intrusive vs. Extrusive

Igneous rocks are generally separated into two groups based on their grain size. Rocks that cool quickly have a small grain size and are called extrusive. Rocks that cool slowly have large crystals and are called intrusive – that is they have cooled underground.

At this station you will see two extrusive rocks and two intrusive rocks.

Station 2:   Rock Textures

Besides the overall grain size, there are several other characteristics of the appearance of rocks that are important in determining the history of the rock. At this station you will examine 5 different textures and make a sketch and description of each. The textures you will see are **phaneritic**, **aphanitic**, **porphyritic**, **vesicular** and **glassy**.

Station 3:   Mineral Content

Besides the grain size, the mineral content of the rock determines its name. At this station you will practice identifying the minerals in several rocks. To identify the mineral in each rock, you should use the mineral properties. You can do basic scratch tests to determine hardness, you can look for the flash of cleavage planes, and you can use color to identify the minerals present. Remember, not all minerals are found in igneous rocks. You will usually only see quartz, potassium feldspar, plagioclase feldspar, biotite, amphibole, pyroxene, muscovite and olivine. There may be some other minerals, but they are usually either rare or the product of weathering or alteration of the original minerals.

Station 4:   Mineral Percentage

At this station you will estimate what percentage of each mineral is present in the rocks. This is often crucial to determining the correct name of the rock. Getting the exact percentage is not critical. Getting as close as possible is. Obviously, correctly identifying the minerals in the rock is important to determining the percentage of those minerals, so you need to do that first.

Several techniques can be used to estimate mineral percentages. You should experiment with each and see which works for you.

Method 1: Charts. Figure 4.1 contains comparative diagrams to estimate mineral percentages. Notice that black minerals tend to dray the eye and most students overestimate them.

Method 2: Random Sampling. Drop a pen or other sharp instrument on the rock 20 times. Each time identify the mineral it falls on. For each count, the mineral is 5%. So if out of 20 times the pointer lands on quartz 3 times, the rock has about 15% quartz. You must point randomly.

Method 3: Mental Imaging. Imagine all of one mineral moved over to one side or corner of the rock. What percentage of the rock would it take up?

Method 4: Ratios. Figure out the rations of each mineral to the others. Use these to figure out the percentage. You should be moderately fluent in basic arithmetic for this. For example, you see that there are equal amounts of quartz and k-feldspar, twice as much plagioclase as k-feldspar, 3 times as much quartz as biotite and equal amounts biotite and amphibole. The only percentages that work are 7.1% biotite, 7.1% amphibole, 21.3% quartz, and k-feldspar, and 46.2% plagioclase.

Rounding errors mean that the numbers add up to 99%. One can add that percent of any of the minerals you think were underestimated.

Name: _____

Please put sample numbers in numerical order.

Table 4.1: Igneous Rock Property Table for Lab 4.1		
	Sample #	**Intrusive or Extrusive?**
STATION 1		

	Sample #	Texture	Sketch and Description
STATION 2		Phaneritic	
		Aphanitic	
		Porphyritic	
		Vesicular	
		Glassy	

Please put sample numbers in numerical order.

	Sample #	Minerals Identified
STATION 3		

	Sample #	Minerals Present	Percentage
STATION 4			

Name: _____

Pre-Lab 4.2

1. What is the extrusive equivalent of granite?

2. What is a phenocryst?

3. If you find a phenocryst of potassium feldspar in a volcanic rock, what possible names could you give to the rock?

Exercise 4.2

Instructions

1. Examine the samples of rocks provided by your instructor. For each of them, with the help of the charts and the table enclosed in this lab manual, fill the blanks in the following table with the answers to all of the applicable questions.

2. In the column of minerals, if the rock is intrusive, write the names and percentage of all the minerals present. If the rock is extrusive and porphyritic, write only the names of the phenocrysts that you can identify and not the percentage. If the rock is extrusive but either aphanitic or glassy leave that cell blank.

3. Finally, using all the information gathered identify the rock by name.

4. In the column of silica content use words such as mafic or felsic.

Name _____

Igneous Rock ID Chart

Sample #	Texture	Intrusive or Extrusive?	If Extrusive: Effusive or Explosive?	If Intrusive: minerals present and percent	Name of rock	Silica Content

Igneous Rock ID Chart

Sample #	Texture	Intrusive or Extrusive?	If Extrusive: Effusive or Explosive?	If Intrusive: minerals present and percent	Name of rock	Silica Content

Name _____

Igneous Rock ID Chart

Sample #	Texture	Intrusive or Extrusive?	If Extrusive: Effusive or Explosive?	If Intrusive: minerals present and percent	Name of rock	Silica Content

Igneous Rock ID Chart

Sample #	Texture	Intrusive or Extrusive?	If Extrusive: Effusive or Explosive?	If Intrusive: minerals present and percent	Name of rock	Silica Content

Name: _____

Questions

1. Using the igneous rocks classification diagram in Figure 4.2 state the percentage (if any) of quartz, K-feldspar, plagioclase feldspar, muscovite, olivine, amphibole, biotite, and pyroxene in a **diorite** with 56% silica.

2. Using your notes and/or book for reference, suggest the volcanic landform (i.e. lava plateau, shield volcano, composite cone, cinder cone, maar, etc.) and the tectonic setting where each of the following rock types might be found.

3.

Rock Name	Volcanic landform	Tectonic setting
Basalt		
Andesite		
Dacite		
Rhyolite		

79

Lab 5

Topographic Maps

All of you probably use maps while driving, hiking, looking at sales data, reading about history, and much more. In this lab we will learn to work with topographic maps, which represent the three dimensional surface of the Earth in two dimensions. If you hike in the mountains, these skills can be lifesaving. If you own a home or property, the topographic data on a map could affect your property and the possible improvements you can or cannot make on it. Furthermore the concept of contour lines is widely used to display all kinds of data on maps, including crime rates, sales, population densities and a multitude of other important data.

In geology, maps are used to document the features of the surface of our planet. We track the changing courses of rivers, the movement of glaciers, even the size of landslides using topographic maps. Sometimes, the types of rocks that are found on the surface are plotted on a topographic map to create a geologic map, one of the fundamental tools of geology.

To Bring to Lab

Pencil   Lab Manual   Ruler   Calculator

Name_____

Pre-Lab 5.1

Read the rules for contouring in the lab. All questions refer to the diagram on this page.

1. What is the elevation of point A?_____

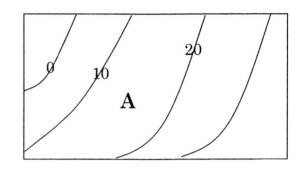

2. What is the contour interval of this map?____

3. Could the unlabeled contour have a value of 20? Explain your answer.

4. If the map has a scale of 1:10000, how far is one centimeter on the map?

83

5.1 Introduction to Topographic Maps and Contours

Introduction

Topographic maps record the elevation of the Earth's surface above mean sea level. This is actually quite a trick since they are reducing our three dimensional world to two dimensions. **Bathymetric maps** work in exactly the same way, except that they record depth of water below sea level.

Scale

Maps should all have a **scale** -- a measurement of how a map distance relates to actual distance. This scale is described in several ways. One way is to place a bar on the bottom of the map on which various distances are marked. The second way is a numerical "proportion scale". This scale is the ratio of map distance to true distance on the ground. In many cases, for instance, you will work with a 1:24000 scale map (read "one to twenty four thousand scale"). On this scale map distances are shrunk by 24,000. One inch on the map is 24,000 inches in the real world, for instance. This is fairly standard scale for topographic maps in the U.S., although 1:62500 and 1:250000 scales are also used to cover larger areas.

Contour Lines

Elevation is represented by **contour lines** -- lines of equal elevation. All points on the same contour are at the same elevation. At the bottom of the map you will find the **contour interval** -- the vertical distance between each adjacent contour. To help you read the map, every 5th contour is bold. To find the elevation at any point, first look to see if it is on a contour. If it is, then it is simply the elevation of that contour. Most points fall between two contours, so we cannot know its exact elevation but only that it is between the values of the contour below and the contour above it. Contours operate by a set of rules. These rules are listed below and will help you in the interpretation of contour maps:

1. Every point on the same contour has the same elevation.

2. Contour lines never split or merge, except at a vertical or overhanging cliff.

85

3. Contour lines are typically drawn at evenly spaced intervals of elevation. This is called the **contour interval**.

4. Contour lines always close on themselves, forming a loop, however this may happen off the edge of the map.

5. The closer together contours are, the steeper the terrain. The more widely spaced they are, the gentler the terrain.

6. Contour lines bend uphill (forming a **V**) when they encounter a stream or the bottom of a valley. The vertex of the **V** points in the uphill direction.

7. Both the peaks of hills and depressions are represented by closed contour lines. Small closed depressions are indicated by hachure marks in order to distinguish them from hills.

8. Contour lines repeat on opposite sides of sloping features like ridges and valleys.

9. Contour lines repeat on opposite sides of depressions and hills except when a depression is on a steep slope.

Elevation, Relief, Height

Elevation is the height above sea level of a particular location. **Relief** is the maximum vertical distance from the highest point in an area to the lowest point in that area. In a region that includes the ocean, the elevation of the highest point is the relief. **Height** is the vertical distance from the bottom of a feature to the top. A demonstration of these differences is that Mt. Everest, at 29,028 ft above sea level, is the highest elevation

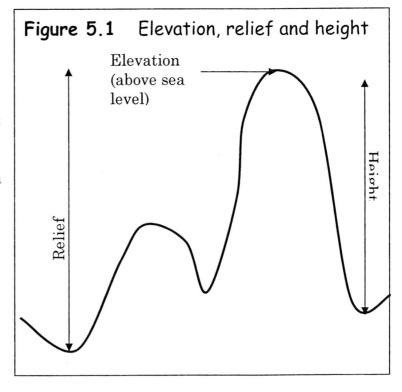

Figure 5.1 Elevation, relief and height

mountain in the world. However Mauna Loa is the one with the greatest height, since from base to top it is more than 30,000 feet tall. Figure 5.1 illustrates the difference between elevation, height and relief.

Profiles

Topographic **profiles** are a graphical way to show the topography along a horizontal line. This is easiest to conceptualize if you imagine taking a big slice out of the Earth and holding it up, looking at it from the side. Another way to imagine this is to think about walking from one end of the line to the other, plotting your elevation versus the distance you have traveled.

To create a topographic profile, follow this step-by-step procedure:

1. Place a piece of paper along the line of the desired profile.

2. Make marks at the ends of the section line and label them (e.g. A-A') so that if it moves, you can put it back in the right place.

3. Make marks and label the elevation on the paper wherever a contour line intersects the profile line. Where the contours are very closely spaced you will not need to determine the elevation of every contour line.

4. Determine the elevation of critical points along the profile. These will include any streams or high points.

5. Get a sheet of graph paper or other grid. Construct a vertical scale by marking off elevations along the vertical axis, making sure that the range of elevations fully encompasses the range of elevations on the map. Use the scale on the map to mark off the horizontal axis scale, so it will have the same horizontal scale.

6. Lay the paper you used before at the bottom of the graph, along the horizontal axis on your graph. Transfer the points to the graph by placing them exactly above the mark on your strip of paper at the corresponding elevation. (Use a ruler to place the points straight.)

7. Connect the dots on the graph with a smoothly drawn line. This is your profile!

Figure 5.2 shows a simple contour map and a profile.

Figure 5.2 Simplified drawing illustrating the method for creating a profile from a contour map

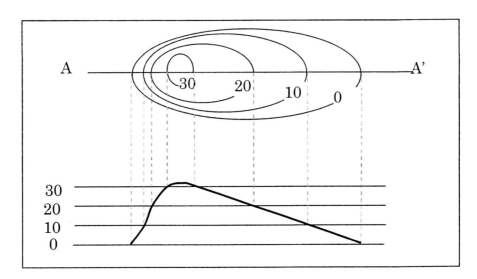

Vertical Exaggeration

Often the vertical and horizontal scales on a profile are not the same, in order to show the data more clearly. Usually, the vertical scale is larger, making the peaks appear higher and the valleys deeper. This is called vertical exaggeration. The vertical exaggeration is determined by dividing the horizontal scale by the vertical scale:

$$\text{Vertical Exaggeration} = \frac{\text{Horizontal Scale}}{\text{Vertical Scale}}$$

Name_____

Exercise 5.1 ——————————————————————

Read over the rules of contour lines and the instructions on how to draw a topographic profile in the introduction. Using the contour map in Figure 5.3, answer the following questions. Turn in this sheet along with the profile in the next page, by the due date assigned by your instructor.

1. On the contour map label the unmarked contours with their elevation in the spaces provided.

2. What is the interval between the bolded contours?_____

3. What is the elevation and name of the highest point on the map? _____

4. What is the elevation and name of the lowest point on the map? _____

5. What is the elevation at point A? _____

6. What is the elevation at point C? _____

7. What is the elevation at point D? _____

8. What is the elevation at point E? _____

9. What is the relief of the area shown on the map?

10. What is the elevation at the bottom of the depression near point C? _____

11. How deep is the depression from the lip to the bottom? _____

12. Circle the area on the map with steepest slope. Put a box around the area with the gentlest slope.

13. How far is it from point C to point E (as the crow flies)? _____

14. Construct a profile along the line X-X' on grid provided on the next page, using an appropriate vertical scale.

18.   What is the vertical scale on your profile in meters per inch?

19.   What is the horizontal scale on your profile in meters per inch?

20.   What is the vertical exaggeration of your profile?

X   X'

Figure 5.3

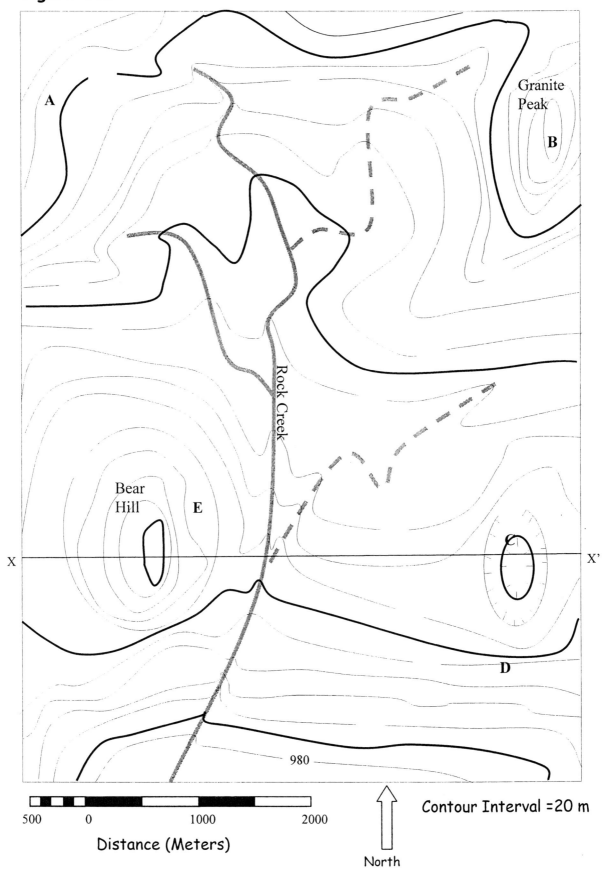

Name_____

Pre-Lab 5.2

Read the explanation in this manual about topographic maps (Section 5.2.)

1.   How many minutes are there in one degree?

2.   Does 1 degree of longitude = 1 degree of latitude? (The answer may not be obvious – explain your answer.)

3.   If a map is at a scale of 1:100000, what distance in the real world will one centimeter on the map represent?

4.   Which are correct representations of a point using the latitude-longitude system?
    (a) 55°N, 186°W
    (b) 145°S, 78°W
    (c) 146°N, 65°S
    (d) 139°S, 35°W
    (e) 37°S, 127°W
    (f) 123°E, 36°N

5.2 Reading Topographic Maps

Introduction

The Map

Maps are traditionally drawn with north up, south down, east to the right and west to the left. All maps, with the exception of a globe, are projections. The surface of the Earth is curved, and maps are flat, so the image must be distorted. The most common projection is the Mercator projection for world maps. These are deceptive in that they significantly distort areas near the poles. Topographic maps are usually a projection that distorts less when working with a small area.

Latitude and Longitude

Latitude and **longitude** are the two axes of a grid system set up on the surface of the Earth so that places (cities, mountains, political boundaries, etc.) can be located in an absolute sense.

Figure 5.4   Latitude and Longitude

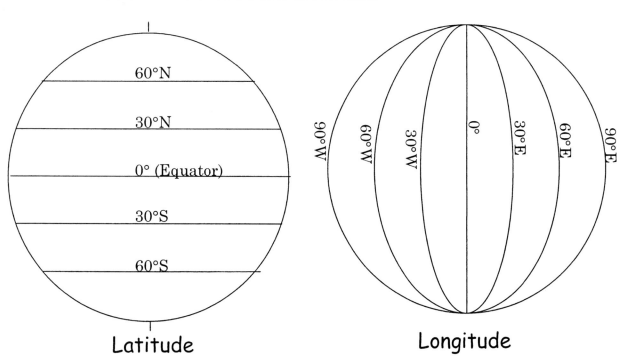

Latitude lines run East-West, and divide the Earth into 180 equal sections, all parallel to the equator. The latitude of the equator is 0 degrees. To the north latitude lines reach 90°N at the North Pole. To the south they go to 90°S (the South Pole). One degree of latitude is always 111 km or roughly 60 nautical miles.

Longitude lines run North-South and divide the Earth into 360 sections (degrees), all joining at the North and South poles. These lines are shaped much like sections in an orange (Figure 5.3). The 0 degree line of longitude runs through Greenwich, England. To the west they go to 180° W and to the east they go to 180° E, a grand total of 360 degrees. At the equator, one degree of longitude is 111 km.

If you walk east-west you are changing longitude, and walking along a latitude line (called a **parallel**). If you are walking north-south, you are changing latitude and walking along a longitude line (called a **meridian**).

Longitudes and latitudes are given in degrees, as stated above. These degrees can be subdivided into minutes and seconds. There are 60 minutes in a degree of latitude or longitude and 60 seconds in a minute.

Very exact locations are easy to give using the latitude-longitude system. However, they must be given in a standard format or misunderstandings or ambiguities are likely. The rules are: latitude first, with degrees symbolized by a °, and, if necessary, minutes and seconds symbolized by ' and " respectively. Then there MUST be either an N or S to tell if one is in the northern or southern hemisphere. Longitude works the same way, but with a W or E to describe the proper hemisphere. Here are some correct examples: 47°34'56"N 3°6'28"E; 34°10'N 118°17' W; 25°S 89°W.

Other Coordinate Systems

There are 2 other ways of marking locations on a map that you should be aware of, but that we will not cover in this lab. The Township-Range system is particular to the US, and is still used for some mining claim and property boundary uses. In this system, the US is divided into 6 mile by 6 mile squares, which are then divided into mile squares. This system is the series of mile by mile squares seen on most U.S.G.S. topographic maps.

The Universal Transverse Mercator system (UTM) is widely used around the world. It is now marked on newer USGS topographic maps. In this system, the world has been divided into 1 km squares, which are all numbered in an east-west fashion. This system is rapidly replacing the latitude-longitude system in most countries outside of the US since it fits well into the metric system.

Name_____

Exercise 5.2 ——————————————

You must SHOW YOUR WORK to get partial credit. PLEASE DO NOT MARK ON THE MAPS!

We will be using the Mount Rainier National Park 1:50000 topographic map.

PART A

1. Look at the bottom of your map. What is the scale? _____

2. At this scale, how far on the map is one mile? _____cm _____in

3. How did you find this out?

4. When was the data from the map collected? (Read from the lower left hand side)

5. Will these maps indicate damage to roads and trails from the 1996 storms?

6. What is the contour interval of this map?_____

7. What is the highest elevation on the map?_____ What is its name?

8. At what type of physical feature will the lowest point on the map be found?

9. What is the lowest elevation on the map?_____ ? Describe where it is.

10. What is the total relief of the mapped area?

11. What is the elevation of the Paradise Visitor's Center? _____

12. How far is it (in kilometers) from the Paradise Visitors Center to Columbia Crest?

Part B

1. What is the latitude and longitude of the corners of the map?

_____ _____ NW corner   _____ _____ NE corner

_____ _____ SW corner   _____ _____ SE corner

2. How many centimeters on the map correspond to 5 minutes of latitude?
   _____cm

3. How far on the map is 1 minute of latitude? _____cm

4. One centimeter is equal to how many seconds of latitude?_____

5. How many centimeters on the map correspond to 5 minutes of longitude?
   _____cm

6. How far on the map is 1 minute of longitude? _____cm

7. One centimeter is equal to how many seconds of longitude?_____

8. What is the <u>exact</u> longitude and latitude of Columbia Crest? You must be within 15 seconds

   _____   _____

9. If you were standing at 46° 54' 39" N 121° 39' 20"W, where are you?

10. Assuming a mudflow from Mount Rainier traveled down the Nisqually River Valley, what areas would be affected? (You may assume the thickness of the mudflow would be about 80 feet) Provide names!

11. Finally, on a piece of graph paper, construct a profile from Needle Rock in the North Mowich Glacier to Anvil Rock, just below Camp Muir.

12. What is the vertical exaggeration of the profile you constructed?

Lab 6

Sands

Introduction

Sand is one of the most ubiquitous, important, and overlooked materials on Earth. In fact, the average American uses almost 4000 kg (8800 pounds) of sand and gravel _every year_. And yet it is rare that a person walking on the beach bends over to take a look at this amazing resource.

Every grain of sediment has a long and usually exciting history. Much of this history can be discovered through close observations of the sediment itself, either through a microscope or a hand-lens. We will do just this, trying to discover the history of several sands collected from throughout the U.S. Can you discover which one was collected on a nearby beach?

For each of the sands you look at you will make several observations: composition, size, roundness, sorting, surface features and any other characteristics you may notice. Then, using these observations you will choose two sands for which you will interpret a history. This history will include several parts: a discussion of the possible **source** (what sort of a rock or area the sand came from), the mode of **transport** (how it was moved from the source to the place where it was collected), and the **duration of the transport**, how long and far I was moved from the source.

To help you describe the sediments, here are some characteristics you want to look at:

Composition

The composition of the sediment is often the hardest thing to determine and one of the most important, because it can tell you what sort of rock the sediment came from. There are obviously as many different compositions as there are rocks and minerals -- almost infinite -- but they can be roughly divided into major groups. Do not worry about identifying each and every grain of sand, but rather look at what type or types predominate.

Rock Fragments -- typically gray dark, rock fragments are composed of more than one mineral, often too small to be seen. These are often volcanic, so a sediment with many of these may indicate a volcano nearby.

99

Biogenic -- these are small pieces of shell or plant debris. They are most commonly found in beach sands from warm climates, although they can be found in many other areas. Biogenic fragments are typically not very durable.

Minerals --There are several mineral grains you might see:

Quartz -- a typically gray or white grain, quartz is glassy. Grains of quartz are commonly eroded from sedimentary rocks. This is one of the most common minerals in sand because it is very durable and hard.

Feldspar -- Typically milky white or pink, this mineral is not as glassy as quartz. Because it is softer than quartz, typically it will not be as common in sediments that have been eroded and transported for many millions of years. Feldspar indicates that an igneous rock (such as granite) was eroded to form the sediment.

Olivine -- Green and glassy mineral, it is very unstable and therefore rare in sediments. Its presence indicates you are near a basaltic volcano, and that the sediment has not been transported a great distance.

Magnetite --Magnetite can be detected by moving a magnet around underneath the paper or tray that the sediment is on. *It is very important that you not touch the magnet to the sediments, as you will remove the magnetite.*

Others-- Other mineral grains may be virtually any color at all, and are frequently dark.

Size

Different sediments have clasts that are different sizes. To determine grain size one measures or estimates the size of the largest, smallest and "average" grain size. Grain size can tell you a lot about a sediment. Big boulders are found closer to mountains, the source of sediments. Fine clay is rarely found in rough mountain streams. On the other hand, it is very rare to find a boulder in the deep ocean. Large boulders are also not found in deposits that traveled by wind, and are more common in landslides.

If you believe a sediment was transported by water, then you can use Hjulstrom's Diagram (Figure 6.1) to estimate how fast the stream or current was moving when that piece of sediment was picked up (eroded) and how fast it was moving when it was deposited. For example, a grain of sand that is 1 mm in size would need a current of about 30 cm/sec (1 ft/sec) to erode it, but it would continue to be transported until the stream started moving slower than 10 cm/sec (3.5 in/sec).

Figure 6.1   Hjulstrom's Diagram

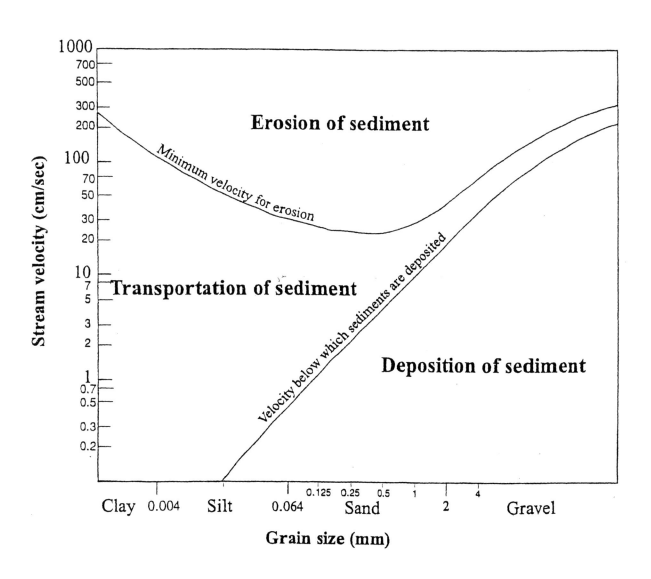

Roundness

Roundness is a measure of how "sharp" the corners are on the grains of sediment. You can use the diagrams given in Figure 6.2 to determine whether the grains are "very angular", "angular", "rounded", etc. Please note that an oblong clast may be very rounded if it has no sharp edges. Typically a grain is more rounded the farther it has been transported from its source, so a round grain may be generalized to have traveled much farther from where it was eroded than an angular grain. Of course composition makes a big difference. Quartz is very durable and takes a long time to become round while olivine is very unstable and will round quickly.

Roundness can even tell you a bit about how something was transported. Typically landslide and glacial deposits are more angular and deposits from water will be more rounded because the clasts grind against each other.

Figure 6.2 Comparison diagrams to determine roundness of a sediment (From "Practical Sedimentology" by D. Lewis, pg. 64, Hutchinson Ross, 1984).

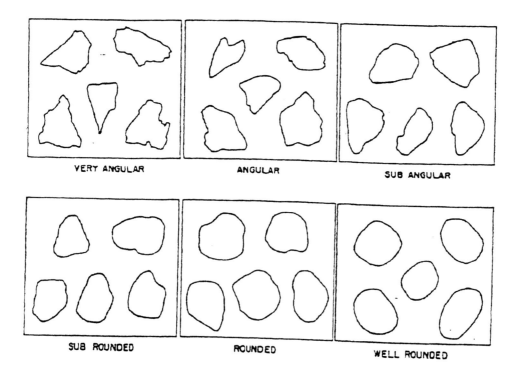

Sorting

Sorting is a measure of how diverse the sizes of the clasts are. Sediment with only one grain size is considered well sorted, while a sediment with both big and small pieces is considered poorly sorted. Use the diagrams provided in figure 6.3 as a guide.

Sorting can often tell us how sediment was transported. If the sediment is well sorted, it was typically moved by water. Very well sorted sediment indicates that it may have been transported by wind. Poorly sorted sediment indicates movement by glaciers, landslides, or mudflows. (You should ask yourself "why?")

Surface Features

Sometimes there are surface features on a clast that will help you figure out some part of its history. The best example of this is the pitting or frosting that comes with transport by wind. Wind-transported sediments are self-sandblasted, and so often have very small pits in the surface or may appear frosted because of the numerous tiny scratches and holes.

Figure 6.3 Diagram for estimating sorting of a sediment
(From "Practical Sedimentology", D. Lewis, pg. 63, Hutchinson Ross, 1984).

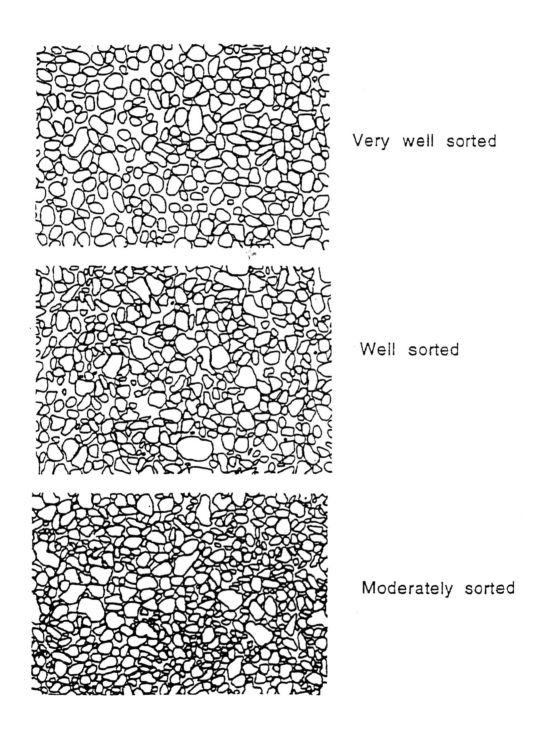

Name_____

Pre-Lab 6

Read over the lab. Answer the following questions:

1. If a sediment is made up of only one size grain it is well _____.

2. What do pits or frosting tell you about a sediment?

3. What modes of transport might cause a sediment to be angular?

4. Which is harder, quartz or feldspar?

Name_____

Exercise 6

Part 1 -- Description

As completely and specifically as you can, describe all five sediments provided. Be sure to use all of the properties discussed in the introduction. Use complete sentences. You may use additional sheets if you need more space.

PLEASE BE VERY CAREFUL NOT TO MIX THE SANDS!!! Several would be virtually impossible to replace in a pure form, and mixing them will only confuse students in the future.

KDS_____

_____

_____

_____

_____

_____

_____

_____

_____

_____

_____

_____

_____

SSP_____

_____

_____

_____

_____

_____

_____

_____

_____

_____

_____

HVP

MSA

HBS

Part 2 -- Interpretation

Now, on a separate piece of paper, interpret as much as you can about the history of two of these sands including the source and transport method, and where you might find this sort of sediment. You will probably not be entirely correct, but that is okay. What is more important than being correct is explaining **EXPLICITLY AND COMPLETELY** why you made the interpretations that you did. A good example would be:

> *"This sediment was transported by a glacier* because <u>it is poorly sorted and not well rounded</u>. While mudflow and landslide deposits are also poorly sorted, they typically are more angular. Also, <u>there were many different grain compositions, including quartz, feldspar, and many different rock fragments</u> which indicates that it was not a landslide which typically is more homogenous. *Therefore the only remaining method of transport is a glacier."*

(I have italicized the interpretations and underlined the descriptions. All the underlined parts should be included in part 1.)

Lab 7

Sedimentary Rocks

Introduction

Sedimentary rocks are often divided into clastic and chemical/biochemical rocks.

Clastic Sedimentary Rocks

Sedimentary rocks composed of clastic particles derived by the physical weathering, erosion, transport, and deposition of pre-existing mineral grains and rocks are said to have **clastic texture** and are called clastic sedimentary rocks. Clastic rocks are classified based on the size of the particles that make up the rock. Table 7.1 shows the Particle Size Classification used for the identification of clastic sedimentary rocks.

A sedimentary rock containing many pebbles and cobbles are known as a **conglomerate** or a **breccia**. In the conglomerate, the larger pebble-size particles are rounded and in a breccia, the pebble-size particles are angular. A rock composed of sand-size particles is called **sandstone**, regardless of the composition of the sand particles. Grains smaller than sand are often too small to be seen without magnification. Silt-size grains can be seen with a hand lens and silt grains are large enough to feel, so **siltstones** feel gritty. Mud-sized particles cannot be seen even with a hand lens and are too small to feel. **Mudstones**, rocks composed of mud particles, often feel smooth. **Shale** is a special type of mudstone that is fissile due to the coplanar orientation of the mineral grains. Shales can be recognized by their strong layering and tendency to break along parallel planes.

The size, shape, sorting, and mineralogy of clastic sediments can provide clues to geologic history in the same way that examining sediment textures (Lab 6) provided clues about transport agents and sedimentary environments. Geologists also examine these features in sedimentary rocks as well as sediments.

Chemical/Biochemical Sedimentary Rocks

Chemical and biochemical sedimentary rocks are classified based on their mineral composition and texture.

Chemical Sedimentary Rocks

Many chemical sedimentary rocks are composed of a single mineral such as quartz or calcite. To identify these rocks, it is often necessary to perform acid, hardness or other tests like those used to identify minerals.

Most chemical sedimentary rocks have a crystalline texture. They are formed by precipitation of minerals out of saturated solutions. Examples of chemical sedimentary rocks are:

Rock salt – a rock composed of the mineral halite (NaCl). Rock salt is salty to taste and has the same hardness as halite.

Chert – a rock composed of quartz (SiO_2). Chert is hard and has conchoidal fracture just like the mineral quartz.

Biochemical Sedimentary Rocks

Biochemical sedimentary rocks are also classified based on mineral composition. Most biochemical sediments have distinctive textures that make them readily identifiable: shell fragments, fossils, or plant debris. Examples of biochemical sedimentary rocks are:

Limestone – a rock composed dominantly of the mineral calcite ($CaCO_3$), which fizzes in dilute acids, such as HCl. **Chalk**, a special type of limestone, is composed of the skeletons of tiny single-celled organisms called coccolithophores. These skeletons, called microfossils, are too small to be seen. Chalk is usually low density, porous, and often has a powdery feel. **Coquina**, another type of limestone, is composed of the fairly large fragments of broken shells and fossil debris of marine organisms cemented together.

Diatomite – a rock composed of microfossils called diatoms. Diatoms create their skeletons of silica. Diatomite is similar in appearance to chalk, but there is an easy way to tell the two apart. Diatomite, composed of silica, does not effervesce in dilute acid.

Coal – a rock made up the compacted remains of plants buried in lagoons, swamps and delta environments. Coal is usually light because if its low density.

Table 7.2 provides a summary for identifying chemical and biochemical sedimentary rocks.

Table 7.1	Particle Size Classification for Clastic Sedimentary Rocks		
	Particle Size (mm)	Name of Clastic Particle	Name of Sedimentary Rock
Coarse ↑	> 256	Boulder	Breccia (angular)
	64 - 256	Cobble	
	2 - 64	Pebble	Conglomerate (rounded)
	0.064 – 2	Sand	Sandstone
	0.004 - 0.064	Silt	Siltstone
↓ Fine	< 0.004	Mud	Mudstone (fissile = Shale)

Table 7.2 Chemical/Biochemical Sedimentary Rocks			
Composition	Rock type	Name of rock	Description
$CaCO_3$	Chemical or Biochemical	Limestone	Various colors (beige, white, gray, pink). Sometimes displays small cavities. Reacts with acid.
	Biochemical	Coquina	Made of small calcareous shells. Reacts with acid.
	Biochemical	Chalk	White or off-white, powdery, very light. Soft. Reacts with acid. Made of microscopic (invisible) remains of microfossils.
SiO_2	Chemical	Chert	Various colors (gray, red, brown, green etc.). Hard, often displays conchoidal fracture.
	Biochemical	Diatomite	White or off-white, powdery, very light. Soft. Does not react with acid. Made of microscopic (invisible) remains of microfossils.
Carbon	Biochemical	Coal	Black. Very light.
NaCl (Halite)	Chemical	Rock salt	Transparent (sometimes dirty on the surface). Salty.

Name _____

Pre-Lab 7

Read Lab 7 and the chapter in your book on sedimentary rocks prior to lab day.
Answer the following questions:

1.   What are the two principle categories of sedimentary rocks?

2.   What characteristic is used to classify (and identify) clastic sedimentary
      rocks?

3.   What characteristic(s) are used to classify (and identify) chemical/biochemical
      sedimentary rocks?

4.   Name two common minerals that make up chemical/biochemical sedimentary
      rocks.

7.1 Textures of Clastic Sedimentary Rocks

Instructions

1. Visit lab stations.
   Your instructor has set up stations in the lab that illustrate different aspects of texture in clastic sedimentary rocks. You may visit the stations in any order. At each station you will:

   a) Review the information on the property that you will be evaluating.

   b) Record the sample numbers, required observations, rock name and the answers to any other questions in the Clastic Sedimentary Rock Identification Chart at the end of Lab 7.1.

2. Turn in your work on the due date given by your instructor.

Station 1:   Grain Size

The average grain size of the particles in a sedimentary rock is a measure of the available energy of the transportation agent and sedimentary environment in which the particles making up the rock accumulated. A **sedimentary environment** is a place on the continent, shoreline, or in ocean where sediments are deposited. Each type of sedimentary environment represents a different set of geologic conditions that determine the types (sizes) of materials that will be deposited there. As a result, the size of the clastic particles that make up sedimentary rocks can give us clues about the geologic conditions that existed where/when the particles were accumulating. For instance, mud particles are so small that they are suspended and transported even by the weakest currents. Mud particles settle and accumulate only in quiet water. Mud particles, therefore, accumulate in places like lake bottoms, tidal flats, or in deep ocean water, where the currents are weak and the mud can slowly settle out. For a discussion of other sedimentary environments, refer to your textbook or your lecture notes.

The rocks at Station 1 are **sandstone**, **siltstone**, and **shale**. Evaluate the grain size and name the rocks. Suggest a possible sedimentary environment for the sediments that make up each of these rocks.

117

Station 2:   Sorting

Sorting measures the uniformity of grain size in a sedimentary rock. Very well sorted rocks are composed of grains that are all one size. Poorly sorted rocks are composed of grains of many different sizes.

The sorting of sediments in rock can provide us clues with how the sediments were transported and deposited. Sediments are transported by water, wind, ice (glaciers), and gravity (mass movements). These are known as **transport agents**. The size of the particles that each transport agent can carry varies based on its ability to lift and entrain particles. For example, wind can only transport particles that are sand-sized or smaller. Even a very strong wind could not pick up gravel or cobble-size pieces (thank goodness!)

Transport agents also vary in their ability to sort sediments according to size as they transport them along. Wind is very effective at sorting sediments because it can only carry sediments of a certain size (leaving the larger particles behind). When the current slows and the sediments are dropped by the wind, they are generally well sorted. You might want to think about the ability of the other transport agents to sort sediments.

The rocks at Station 2 are **sandstone**, **breccia**, and **conglomerate**. Evaluate the sorting of the sediments in the rocks and name them. Based on the sorting, suggest a possible transport agent for the sediments in the rock.

Station 3:   Grain Roundness

The impact between particles during sediment transport results in **abrasion**. Abrasion causes the particles to become smaller as they break apart from forceful collisions. Continuous impacts wear off and smooth sharp edges of the particles causing them to become **rounded**.

The degree of grain roundness is related to the mineralogy of the particles (how "resistant" it is), the shape and size of the original particle, and the distance that the particle has been transported. Sand grains are less likely to break into several pieces, but tend to become rounded over long distances. Thus, a generalization (not always true) can be made that the degree of roundness of the particles in a rock can be an indicator of how far the particles have been transported from their source area. Angular particles have probably not been transported far from their source; rounded particles have probably been transported farther from their source.

The rocks at Station 3 are **conglomerate** and **breccia**. Examine the degree of rounding of the pebbles in the rocks and identify them by name.

Station 4:   Mineralogy and Maturity of Sandstones

The mineralogy of a clastic rock depends upon the mineralogy of the source area and the effects of differential abrasion on the minerals during the sediment transport process. Some minerals, such as quartz are *chemically resistant* and, due to their hardness and lack of cleavage, physically "tough". We call them **resistant minerals**. Others, such as feldspars, amphiboles, and pyroxenes are susceptible to chemical weathering and/or may be soft or easily cleaved. These minerals are quickly worn away and are **non-resistant minerals**.

During extended episodes of transport, the ratio of resistant to non-resistant minerals will increase, the sediment will become better sorted and the grains will become more rounded. A sedimentary rock with a high resistant/non-resistant ratio, high rounding, and excellent sorting is said to be of high maturity. Sedimentary rocks may also be of moderate or low maturity based on the particle sorting, rounding, and the presence or lack of matrix material (mud) between the mineral grains.

Sandstones fall into three groups on the basis of their mineralogy and texture:

> **Quartz Arenite.** Quartz arenites are composed of almost entirely well sorted and rounded quartz sand. These pure quartz sands can result from extensive weathering during transport that removes almost everything but quartz, the most stable mineral.
>
> **Arkose**. Arkoses usually contain up to 25% potassium feldspar and the sand grains tend to be poorly rounded and less well sorted than quartz arenites. They are usually derived from the rapid physical weathering of granitic and metamorphic rocks.
>
> **Graywacke**. Graywackes are a mixture of small rock fragments and angular grains of quartz and feldspars. A fine-grained clay matrix usually surrounds the sand grains. The clay matrix often comes from the chemically alteration and compaction of some of the non-resistant minerals in the sediments.

The rocks at Station 4 are **quartz arenite**, **arkose**, and **graywacke**. Examine the rock samples with a microscope or hand lens, note the presence of distinctive minerals, rock fragments, or matrix and name the rock. Based on the descriptions, determine whether the rock maturity is high, moderate, or low.

Name _____

Please put the samples at each station in numerical order.

Clastic Sedimentary Rock ID Chart			
	Sample #	Grain Size	Rock Name	Possible Sedimentary Environment
STATION 1				

	Sample #	Degree of Sorting	Rock Name	Possible Transport Agent
STATION 2				

	Sample #	Round or Angular Grains?	Rock Name
STATION 3			

	Sample #	Minerals present	Rock Name	Maturity (high, moderate, low)
STATION 4				

121

7.2 Identification of Sedimentary Rocks

Instructions

1. Use Table 7.1 and Table 7.2 to identify clastic and chemical/biochemical sedimentary rocks.

   Your instructor has provided a number of samples of both clastic and chemical/biochemical sedimentary rocks. You may use microscopes, hand lenses, acid bottles, glass plates, or any other equipment you need to perform any tests or observations.

2. Fill the blanks in the Sedimentary Rock Descriptions/Identification Chart with sample number, all the applicable observations and descriptions, and the rock name.

3. Turn in your work on the due date given by the instructor.

Name _____

Sedimentary Rock Descriptions/Identification Chart			
Sample #	Rock Type: (Clastic, Chemical, or Biochemical)	Chemical Composition (for Chemical and Biochemical Rocks only)	Rock Name
	Brief Description: may include any applicable characteristics: grain size, texture, mineral composition, color, hardness, etc.)		
	Brief Description: may include any applicable characteristics: grain size, texture, mineral composition, color, hardness, etc.)		
	Brief Description: may include any applicable characteristics: grain size, texture, mineral composition, color, hardness, etc.)		
	Brief Description: may include any applicable characteristics: grain size, texture, mineral composition, color, hardness, etc.)		

Name _____

Sedimentary Rock Descriptions/Identification Chart			
Sample #	Rock Type: (Clastic, Chemical, or Biochemical)	Chemical Composition (for Chemical and Biochemical Rocks only)	Rock Name
	Brief Description: may include any applicable characteristics: grain size, texture, mineral composition, color, hardness, etc.)		
	Brief Description: may include any applicable characteristics: grain size, texture, mineral composition, color, hardness, etc.)		
	Brief Description: may include any applicable characteristics: grain size, texture, mineral composition, color, hardness, etc.)		
	Brief Description: may include any applicable characteristics: grain size, texture, mineral composition, color, hardness, etc.)		

Name _____

Sedimentary Rock Descriptions/Identification Chart			
Sample #	Rock Type: (Clastic, Chemical, or Biochemical)	Chemical Composition (for Chemical and Biochemical Rocks only)	Rock Name
	Brief Description: may include any applicable characteristics: grain size, texture, mineral composition, color, hardness, etc.)		
	Brief Description: may include any applicable characteristics: grain size, texture, mineral composition, color, hardness, etc.)		
	Brief Description: may include any applicable characteristics: grain size, texture, mineral composition, color, hardness, etc.)		
	Brief Description: may include any applicable characteristics: grain size, texture, mineral composition, color, hardness, etc.)		

Name _____

Sedimentary Rock Descriptions/Identification Chart			
Sample #	Rock Type: (Clastic, Chemical, or Biochemical)	Chemical Composition (for Chemical and Biochemical Rocks only)	Rock Name
	Brief Description: may include any applicable characteristics: grain size, texture, mineral composition, color, hardness, etc.)		
	Brief Description: may include any applicable characteristics: grain size, texture, mineral composition, color, hardness, etc.)		
	Brief Description: may include any applicable characteristics: grain size, texture, mineral composition, color, hardness, etc.)		
	Brief Description: may include any applicable characteristics: grain size, texture, mineral composition, color, hardness, etc.)		

Lab 8

Metamorphic Rocks

Introduction

In this lab you will learn how to identify some of the more common metamorphic rocks. You will first learn how to distinguish between foliated and non-foliated metamorphic rocks and then you will identify the most common metamorphic rocks.

Metamorphic rocks form when a rock is buried deep into the Earth (several kilometers below the surface) and it is subject to increasing **pressure** and **heat**. These conditions cause changes in the minerals of the original rock, which will be more significant as the amount of pressure and heat increases. Changes include reorientation of the minerals, partial dissolution, migration, recrystallization, and formation of new metamorphic minerals. The original igneous or sedimentary rock is called the **parent rock**. If metamorphism has been extensive it may not be possible to recognize the parent rock. When it is possible to determine that the parent rock was sedimentary, we refer to the metamorphic rocks as meta-sedimentary.

Foliated Metamorphic Rocks

Foliated metamorphic rocks generally form when a rock is caught in a convergent plate boundary and therefore subject to **directed pressure**. This type of metamorphism is called **regional metamorphism**. As directed pressure and temperature increase the **metamorphic grade** increases and the changes in the rock become more extreme. At low metamorphic grade, directed pressure causes the elongated minerals present in the rock to align in the direction of least stress. This internal rearrangement of the minerals is displayed externally by a layering appearance, called **foliation**. Higher metamorphic grades may cause partial dissolution and recrystallization of minerals in a preferred orientation or even the break down of the original minerals and the formation of entirely different minerals.

Foliated meta-sedimentary rocks may form from a common parent rock in a continuous sequence as metamorphic grade increases. For example, the sedimentary rock shale may turn into **gneiss** by going through a continuous

129

series of changes, as the metamorphic grade increases. The sequence of changes is outlined as follows:

- Relatively low temperature and directed pressure will cause the clay minerals in the shale to orient parallel to each other and perpendicular to the direction of applied pressure and will result in a subtle foliation of the rock. This first change will turn the shale into a **slate**. A slate is a smooth rock with **slaty cleavage** (tendency to break along thin layers). If struck against a hard surface slate makes a clinking sound. Slate is a rock commonly used as roofing material. The color of slate depends on the original composition of the shale that it comes from. It can be black if it is rich in carbon, red if rich in iron, green if rich in a mineral called chlorite.

- If the pressure and temperature continue to increase, mica minerals (biotite and muscovite) will begin to recrystallize parallel to one another and grow in size. The rock that forms in these conditions is a **phyllite**. The grain size of a phyllite is still too fine to be seen but the mica minerals give the rock a characteristic shiny appearance called **sheen**. Another distinctive characteristic of phyllite is a well-developed foliation, which is displayed by thin, sometimes curved, layering.

- At even greater pressure and temperature, the mica minerals continue to grow and become visible to the naked eye. Other minerals such as garnet or epidote may form in these conditions. The resulting rock is **schist**, a medium to coarse-grained rock whose extensive foliation is sometime called **schistosity**.

  Schist may also form from an igneous mafic rock such as basalt. This type of schist would normally contain green minerals such as chlorite and epidote and it is called **greenschist**.

- Even higher temperature result in more extreme changes. The muscovite breaks down. Ions dissolve from the existing minerals and recrystallize into new minerals more stable at high pressure and temperature. These minerals segregate into layers of similar composition. The felsic minerals (quartz, K-feldspar, plagioclase) segregate in light-colored bands and the mafic (biotite, amphibole and pyroxene) into dark-colored bands. The resulting coarse-grained foliated rock, displaying alternating black and white bands is called **gneiss** (pronounced "nice").

  Gneiss may also form from felsic igneous rocks such as granite or granodiorite and it may be impossible in some cases to determine the composition of the parent rock.

- If gneiss is subject to even higher pressure and temperature the felsic, low-melting temperature minerals begin to melt. If the rock is let cooled from this stage will display a complex pattern of dark bands and contorted white bands. We would call such a rock a **migmatite**. If the temperature further increases the rock will melt completely and turn into magma.

Nonfoliated Metamorphic Rocks

Non foliated metamorphic rocks form when an igneous or sedimentary rock is either deeply buried within the Earth or comes in contact with magma. In the first case the **confining pressure** of the surrounding rocks is the main cause of **burial** metamorphism. If instead, the rock comes in contact with magma, the intense heat is the main cause of **contact or thermal** metamorphism. In either case, minerals will become unstable in the new surrounding environment and will either recrystallize into more densely packed crystalline structures or into new metamorphic minerals.

When sedimentary rocks consisting mainly of one mineral, are subject to metamorphism, they recrystallize into a non-foliated coarse-grained metamorphic rock with the same composition of the parent rock but with a more dense (and therefore more stable) crystalline structure. For example, pure limestone subject to deep burial or contact metamorphism would turn into a **marble**. Quartz-rich sandstone or chert would turn into a **quartzite**.

Table 8 summarizes the information necessary to identify and describe the most common metamorphic rocks.

Table 8.1 Foliated Metamorphic Rock Identification Guide

Texture	Parent rock	Metamorphic conditions	Metamorphic grade	Grain size of minerals	Possible minerals	Rock name	Distinctive characteristics
Foliated	Shale	Increasing directed pressure and temperature	Low	Very fine (microscopic; invisible to the naked eye)	Clay minerals, micas, chlorite	**SLATE**	Slaty cleavage; clinking by percussion; smooth
	Shale		Low-intermediate	Fine (barely visible to the naked eye)	Micas, chlorite	**PHYLLITE**	Sheen
	Shale, basalt, sandstone		Intermediate-high	Medium (visible)	Micas, chlorite, garnet, epidote, amphibole, talc	**SCHIST**	Schistosity
	Shale, felsic igneous rock, sandstone		High	Coarse	Quartz, K-feldspar, plagioclase, micas, pyroxene, amphibole	**GNEISS**	Black/white banding from alternating layers of felsic and mafic minerals
	Mafic minerals	Moderate pressure and temperature	Moderate	Moderate	Amphibole and Plagioclase	**AMPHIBOLITE**	Usually weakly foliated or non-foliated; salt and pepper appearance
	Basalt	Hydrothermal	Variable	Fine	Serpentine	**SERPENTINITE**	Soapy feel; Green streaky appearance

Table 8.1 (continued) Non-foliated Metamorphic Rock Identification Guide

Texture	Parent rock	Metamorphic conditions	Metamorphic grade	Grain size of minerals	Possible minerals	Rock name	Distinctive characteristics
Non-foliated	Pure limestone	Contact with magma or confining pressure from deep burial	Variable	Fine to coarse	Calcite	MARBLE	Can be scratched with a steel blade; reacts with acid
	Quartz-rich sandstone		Variable	Fine to coarse	Quartz	QUARTZITE	Very hard; does not react with acid
	Basalt or gabbro	Very high pressure	High	Coarse	Garnet and Amphibole	ECLOGITE	Often has a green and red color (Christmas rock)
	Carbonate	High temperature		Coarse	Wollastonite and Garnet	SKARN	White with distinctive garnets
	Mafic minerals	Moderate pressure and temperature	Moderate	Moderate	Amphibole and Plagioclase	AMPHIBOLITE	May be weakly foliated; salt and pepper appearance
	Mudstone or shale	Contact or Thermal	Variable	Fine	Quartz and many others	HORNFELS	Dark gray or black with some white streaks; very hard
	Mafic rocks	Regional metamorphism	Low	Fine	Epidote and Chlorite	GREENSTONE	Greenish color
	Basalt	Hydrothermal	Variable	Fine	Serpentine	SERPENTINITE	Soapy feel; Green streaky appearance

Name _____

Pre-Lab 8

1. List four (4) minerals typically found in metamorphic minerals.

2. If the sedimentary rock shale is subjected to progressively increasing grades of metamorphism, a sequence of new rocks will form from that single parent rock. List the sequence of rocks formed from shale during increasing pressure and temperature conditions.

3. In your own words, define metamorphic foliation and explain what causes it to develop.

4. Quartzite and Marble look similar but are composed of two very different minerals. Identify one mineral test that you could use to tell them apart.

135

Exercise 8

Instructions

1. Examine the samples of rocks provided by your instructor.

2. For each of them, using the information in this lab manual, fill the blanks in the table on the next page with the answers to all of the applicable questions.

3. Finally, using the information gathered identify the rock by name.

NAME _____

Please list samples in numerical order.

Sample #	Texture: Foliated or non-foliated?	Visible Minerals	Texture description	Name of rock

NAME _____

Please list samples in numerical order.

Sample #	Texture: Foliated or non-foliated?	Visible Minerals	Texture description	Name of rock

Name _____

Questions

A **metamorphic facies** refers to a group of metamorphic minerals formed under a specific range of pressures and temperatures. The conditions under which metamorphic facies form can be seen in Figure 8.1, a Metamorphic Facies Diagram. Each area on the graph is labeled with a facies name (Hornfels, Blueschist, Eclogite, etc.). The area represents the range of pressures and temperatures in which minerals of that facies form.

Also plotted in Figure 8.1 are different geothermal gradient paths that show how temperature changes with pressure and depth below the surface in specific geologic settings, such in subduction zones, beneath magmatic arcs, near igneous intrusions, and with deep burial away from plate boundaries.

Figure 8.1 Metamorphic Facies Diagram

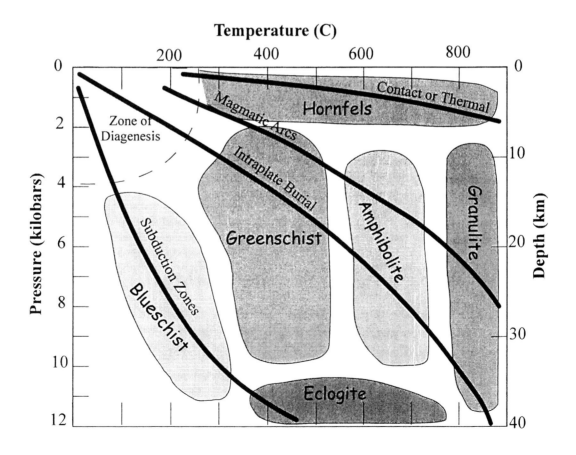

1. To which facies would a mineral that is stable at 400°C and 4 kilobars of pressure belong?

2. What range of temperature and pressure conditions creates the Eclogite facies?

3. Suppose basalt magma is squeezed upward, very near the Earth's surface. It cools to form a shallow igneous intrusion. The rocks around the intrusion have been metamorphosed. Which facies of metamorphism might we expect?

4. In the accretionary wedge near an ocean trench, the rocks have been metamorphosed. Consider the pressures and temperature conditions you might find in this zone. This produces which facies of metamorphism?

5. Beneath the Cascade Mountains, magmas fueled by the subduction of the Juan de Fuca Plate, rise toward the surface. What would be the temperature and pressure conditions for a rock buried 15 km deep below the mountains? This produces which facies of metamorphism?

6. In the Gulf of Mexico, deposits of sediments by the Mississippi River accumulate and are buried to great depths. What would be the temperature and pressure conditions for a rock buried 15 km deep below the Gulf? This produces which facies of metamorphism?

Lab 9

Earthquakes

Introduction

As plates move, **stress** builds up at their margins, like a stick that you try to break across your knee. Eventually, the stress overcomes the strength of the rocks and they break and shift along **faults**. The shifting releases elastic energy that had accumulated, in the form of **seismic waves**. Seismic waves move outward in all directions from the **focus** of the earthquake, the exact place where the breaking and shifting of rocks begins. When seismic waves reach the Earth's surface they cause the shaking of the ground that people feel – an **earthquake**.

A **seismograph** is a machine that can detect and record on paper the shaking produced by seismic waves. This result in a graph made of a series of characteristic squiggles called a **seismogram**. Careful analysis of seismograms shows that there are several different types of waves produced in an earthquake. The first two to be recorded (and the ones we will use in this lab) are:

P-waves – Primary or pressure waves. These are the fastest traveling waves and the first recorded on a seismogram. People often report that they feel like a large truck going by, a distant rumble or even a blast. They do not cause much shaking.

S-waves – Secondary or shear waves. These are slower than P waves and are the second waves to arrive at the seismograph. They cause a lot of shaking and often do most of the damage in an earthquake.

To Bring to Lab

| Pencil | Lab Manual | Drafting Compass | Ruler |

143

Name_____

Pre-Lab 9

Read the introduction to the lab.

1. Using Figure 9.2, answer the following question. An earthquake occurs 200 kilometers from a seismograph. If the earthquake occurred at 7:13:56, what time will the P and S waves arrive at the seismograph?

    P Wave:_____

    S-Wave:_____

2. Look at the seismogram from Des Moines in Figure 9.1. How far is this station from the epicenter?

3. What is the magnitude of the earthquake felt in Des Moines that is shown in Figure 9.1? Use Figure 9.5.

Exercise 9

You will be analyzing a seismogram to determine some basic information about the earthquake that produced it, including the location of the epicenter and Richter Magnitude. But before you get started, here is an example:

Example

An example of a seismogram recorded at a seismograph in Des Moines, WA is shown in Figure 9.1. Use the first significant displacement from the horizontal line to determine the arrival time. From this seismogram you can see that the P-wave arrived at 8:00:12 (12 seconds after 8 a.m.), and the S waves arrived at 8:00:24.

Figure 9.1 Seismogram from Des Moines, WA

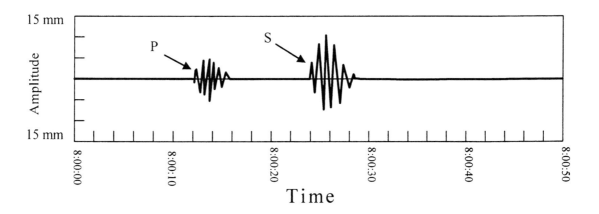

What the seismogram does not indicate is where the epicenter was and when the earthquake occurred. (The P-wave arrived in Des Moines at 8:00:12, but it started sometime before, when the earthquake occurred.)

Distance from Epicenter

To discover how far away the earthquake epicenter was, and when the earthquake occurred, you will need to use the time-travel graph, Figure 9.2. This shows the time required for the seismic waves to travel a particular distance. Because P-waves travel faster than S-waves, as the waves travel away from the epicenter, the time-lag between the two waves increases. You can use this time lag to determine the distance from the epicenter.

In order to determine the distance to the epicenter, you will need to determine the time-lag between the P-and S-wave arrival times, to do this:

1. Determine the arrival time of the P and S waves.

2. Subtract the P-wave arrival time from the S-wave arrival time.

3. Place a piece of paper along the time axis of the time travel graph (figure 9.2) and mark off the length corresponding to the time-lag you just determined

4. Slide the piece of paper along the P and S-wave time travel curves until the two marks line up. Be sure to keep the paper vertical!

5. Where they line up, read straight down to find the distance to the epicenter.

6. Practice on the Des Moines seismogram (Figure 9.1). You should get a time-lag of 12-seconds, which corresponds to a distance of about 100 km.

Location of Epicenter

To find the location of the epicenter, three seismographs are needed. Each one determines the distance from the epicenter. On a map you draw three circles around the locations of the seismographs with radii that are the appropriate distance. There should be one place where the circles intersect, the epicenter of the quake. In some cases small errors make the circles not intersect exactly, but a point closest to the three is a good estimation.

Origin Time

Once you know how far away the earthquake was, you can find the time the earthquake occurred. By looking at Figure 9.2 you can determine how long it takes a P-wave to travel any distance. Looking at the P-wave curve, you can read on the time axis that it will take a P wave 12 seconds to travel the 100 km from the epicenter. Because the P-wave arrived 12 seconds after 8 a.m., the earthquake must have happened at 8:00:00.

The Magnitude of the Earthquake

Once the distance to the earthquake is found, the amplitude of the seismic waves can be used to determine the Richter magnitude of the earthquake. A diagram called a **nomogram** (Figure 9.5) correlates the distance, amplitude and the magnitude of the earthquake. To determine the magnitude follow these steps:

1. On the left line of figure 9.5 find the distance to the epicenter from the seismogram you were using.

2. Measure the amplitude of the S-waves from the seismogram. The amplitude is the total vertical displacement from the center line to the lowest point or the highest point of the seismogram's "squiggle."

3. On the right side of the nomogram, make a mark at the correct amplitude.

4. Connect the distance with the amplitude with a line. It will pass through the middle line.

5. Where the line crosses the middle scale is the Richter magnitude.

Practice with the Des Moines seismogram, as part of your Pre-Lab.

Figure 9.2 Time-travel graph for P and S waves

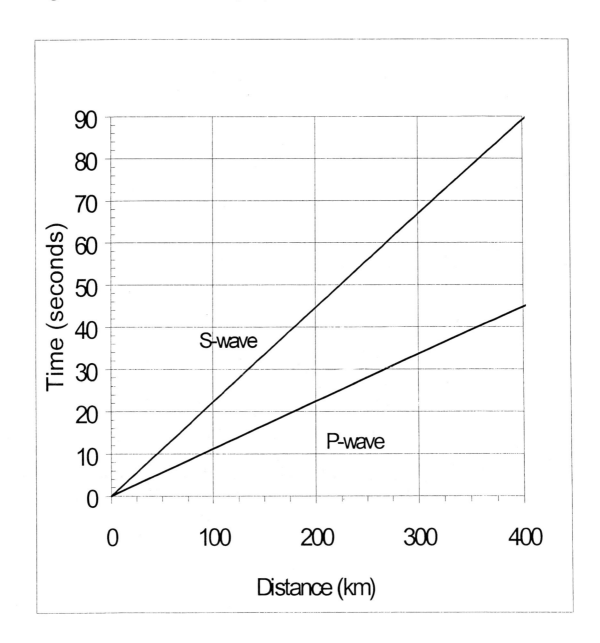

Name_____

Instructions

Complete the questions on this page and submit them with figure 9.4 by the due date given by your instructor.

1. Look at the three seismograms in Figure 9.3, and fill in the table for each seismic station.

	P-wave arrival time	S-wave arrival time	Lag time between P and S-waves	Amplitude
Ellensburg				
Portland				
Bellingham				

2. Using the time-travel graph determine the distance from each seismograph to the epicenter.

   Ellensburg _____km

   Portland _____km

   Bellingham _____km

3. Using a drafting compass, determine the location of the epicenter on the map of Washington (figure 9.4).

4. At what time did the earthquake occur? Show your work.

5. What was the Richter Magnitude of the earthquake? Use the amplitude from the nearest station.

6. If the shaking was 10 times worse, what would trhe Richter Magnitude be?

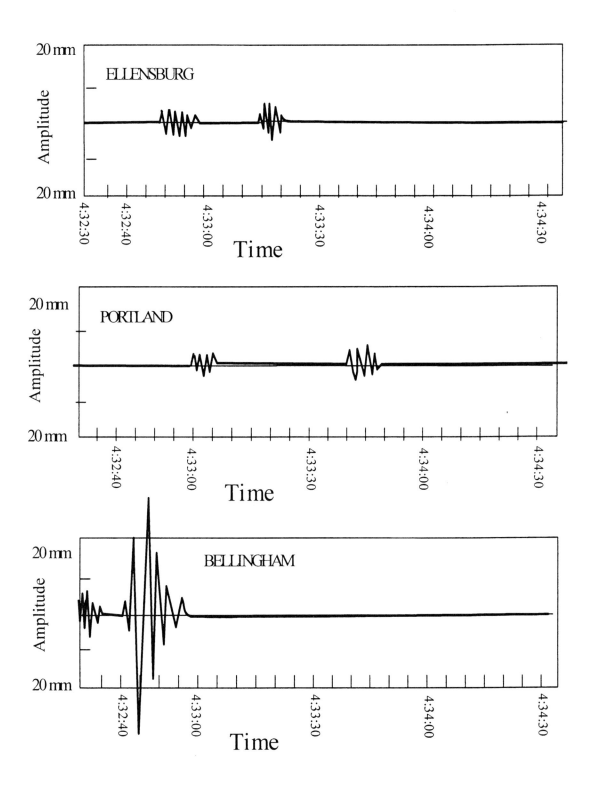

Figure 9.3 Seismograms for an Earthquake in the Pacific Northwest.

Figure 9.4 Map of seismograph locations.

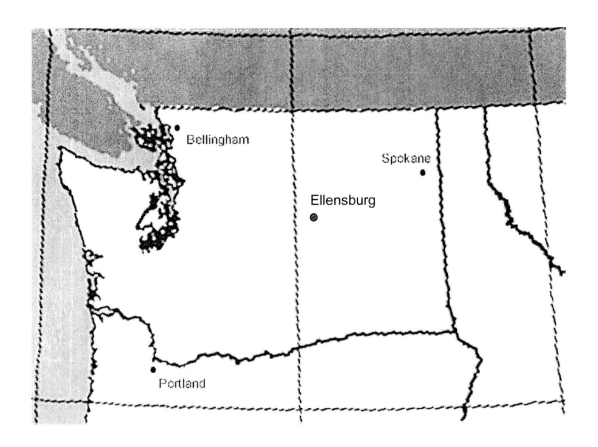

Figure 9.5 Richter nomogram for determining magnitude of an earthquake

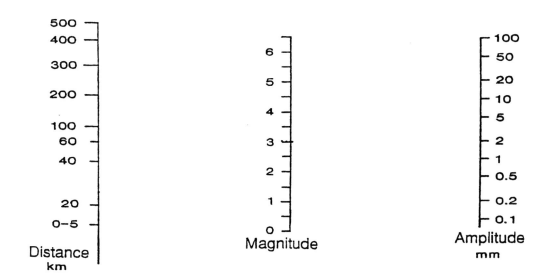

Lab 10

Density and Common Earth Materials

Introduction

The Earth is made of many materials, each with different densities. Density is a function of the atomic weight of the atoms that make up the material (some atoms are heavier than others) and how closely together the atoms are packed (tight packing leads to higher density).

The density of an object or material can be calculated: **Density = Mass ÷ Volume**

We can easily determine the mass of an object by using a mass balance or scale. The volume of material can also be easily calculated, provided it is a perfectly formed geometric shape, like a square, cylinder, or sphere. It is more difficult to calculate the volume of irregularly shaped objects. Do you know what your volume is?

Fortunately, a Greek mathematician and inventor names Archimedes (287-212 B.C.) discovered a convenient way to calculate the volume of any object, not matter what its shape. Legend has it that, while bathing, Archimedes discovered the Principle of Buoyancy. Buoyancy refers the upward force acting on an object immersed or floating in a fluid, like water. Archimedes realized that when you place a solid object in water, the object displaces an amount of water equal to its own volume! All you have to do to calculate an object's volume then is to dunk it in water and measure the amount of water that is displaced!

In this lab, you will measure the densities of four different rocks. The four rocks have been selected because their compositions and/or densities are representative of the different layers of the planet.

Granite: The pink and white minerals in this rock are feldspars. It is a common rock in the <u>continental crust</u> and a good approximation for its density.
Basalt: This rock is black. Most of the <u>oceanic crust</u> is made up of basalt.
Peridotite: This rock contains a green mineral called olivine. It is the best representation for the composition and density of the <u>mantle</u>.
Magnetite: This rock is composed mostly of iron and will be used as a proxy for the composition of the <u>core</u>. (Note: The core is not made of magnetite. We have no samples of the Earth's core, but we do suspect that it is composed mostly of iron.)

NAME _____

Pre-Lab 10

Please answer the following questions. **Always remember to show math work and include units.** Since Mass is usually measured in grams and volume in cubic centimeters, density is usually expressed as grams per cubic centimeters (g/cm^3).

Density = Mass/Volume

1. If a sample has a volume of 13 cm³ and a mass of 30 grams, what is its density?

2. You will be measuring the volume of a rock by immersing it in water and recording the change in water volume. If you had a beaker with water up to the 123 cm³ line and then, after putting a rock in, the water level rose to the 345 cm³ line, what would the volume of the rock be?

3. A rock has a mass of 450 grams. If the volume of the rock is the same as found in question #2, what is the density of the rock?

4. What is the percentage error in your measurement if you misread the beaker and determined the volume to be 224 cm³?

Hint: $$\% \text{ Error } = \frac{\text{Accepted Value - Measured Value}}{\text{Accepted Value}} \times 100$$

159

Exercise 10   Density and Common Earth Materials

Instructions

1. Divide into groups of three to four people.

2. Each group should select one of each of the four rock samples, a overflow beaker, and a graduated cylinder.

3. Record the name of the samples in Density Calculation Table.

4. Using the mass balance, determine the mass of each sample and record it in the Density Calculation Table.

5. Fill the overflow cup until water comes out the spout. Let it drip. Take an empty graduated cylinder and put it under the spout. GENTLY place the rock sample in the overflow cup and catch all of the water that comes out. Measure the volume of water – this is the rock volume.

6. Repeat the volume measure (step #5) another two times with each sample, recording all results.

7. Take the middle of the three trials and use it to calculate the density of each sample.

8. Answer the questions.

NAME _____

Density Calculation Table for Lab 10

Rock Name	Mass in grams (g)	Volume Trials	(A) Water Volume Start (cm³)	(B) Water Volume Finish (cm³)	(B – A) Rock volume (cm³)	Density (g/cm³)
		1				
		2				
		3				
			Average Rock Volume =			
		1				
		2				
		3				
			Average Rock Volume =			
		1				
		2				
		3				
			Average Rock Volume =			
		1				
		2				
		3				
			Average Rock Volume =			

Name _____

Questions

1. The Earth has a mass of 5.976 x 10^{27} g and a volume of 1.083 x 10^{27} cm^3. Calculate the density of the whole Earth.

2. Compare the density of the Earth to the density of the Earth's crust. Do you think the Earth's interior is made of the same material as the Earth's crust? Based on their difference in density, do you think the Earth's interior is made of the same material? Why or why not?

3. What is the relationship between density and the layers of the Earth?

4. What is the order of four rocks in terms of their densities?

    Least dense _____ _____ _____ _____ Most Dense

5. Do your calculations match your expectations? If so, why? If not, suggest a possible explanation.

6. Which is denser, the continental crust or the oceanic crust?

7. The Earth's crust (and lithosphere) is in gravitational equilibrium with the underlying mantle. The mantle under the crust is solid, but the rock there is heat weakened and can "flow". From this information, suggest a reason why the oceanic crust is everywhere at a lower elevation than the continental crust?

8. If you splashed water out of the cylinder while putting the rock into it, how would this affect the result of your **volume** calculation? (i.e., would the volume of the rock be higher or lower?)

9. If you splashed water out of the cylinder while putting the rock into it, how would this affect the result of your **density** calculation? (i.e., would the density be higher or lower?)

Name _____

10. Assume that the average volumes you determined for the four rocks are the true accepted values. If another person misread the graduated cylinder and determined that the volume of each rock was 1 cm^3 less than the accepted value, what would be the percentage error in his/her calculations?

Remember: % Error = $\dfrac{\text{Accepted Value - Measured Value}}{\text{Accepted Value}}$ x 100

Rock	Accepted Value (cm^3)	Measured Value (cm^3)	% Error
Granite			
Basalt			
Peridotite			
Magnetite			

11. Is the % error greater for rocks with a large volume or rocks with a small volume? What might this mean for your own measurements?